HEAVY METALS AND ENVIRONMENT

MOHAMMAD ATHAR
Former Professor and Head
Department of Medical Elementology and Toxicology and
Dean
Faculty of Science, Jamia Hamdard
Hamdard University, New Delhi

SHASHI B VOHORA
Former Faculty
Department of Medical Elementology and Toxicology
Hamdard University, New Delhi

NEW AGE REFERENCE

An Imprint of

NEW AGE INTERNATIONAL (P) LIMITED, PUBLISHERS
LONDON • NEW DELHI • NAIROBI
Bangalore • Chennai • Cochin • Guwahati • Hyderabad • Kolkata • Lucknow • Mumbai
Visit us at www.newagepublishers.com

Published by New Age International (P) Ltd., Publishers
First Edition: 1995
Reprint: 2017

GLOBAL OFFICES

- **New Delhi** **NEW AGE INTERNATIONAL (P) LIMITED, PUBLISHERS**
7/30 A, Daryaganj, New Delhi-110002, (INDIA)
Tel.: (011) 23253771, 23253472, **Telefax:** 23267437, 43551305
E-mail: contactus@newagepublishers.com • Visit us at www.newagepublishers.com
- **London** **NEW AGE INTERNATIONAL (UK) LTD.**
27 Old Gloucester Street, London, WC1N 3AX, UK
E-mail: info@newacademicscience.co.uk • Visit us at www.newacademicscience.co.uk
- **Nairobi** **NEW AGE GOLDEN (EAST AFRICA) LTD.**
Ground Floor, Westlands Arcade, Chiromo Road (Next to Naivas Supermarket)
Westlands, Nairobi, KENYA, **Tel.:** 00-254-713848772, 00-254-725700286
E-mail: kenya@newagepublishers.com

BRANCHES

- **Bangalore** 37/10, 8th Cross (Near Hanuman Temple), Azad Nagar, Chamarajpet, Bangalore- 560 018
Tel.: (080) 26756823, **Telefax:** 26756820, **E-mail: bangalore@newagepublishers.com**
- **Chennai** 26, Damodaran Street, T. Nagar, Chennai-600 017, **Tel.:** (044) 24353401
Telefax: 24351463, **E-mail: chennai@newagepublishers.com**
- **Cochin** CC-39/1016, Carrier Station Road, Ernakulam South, Cochin-682 016
Tel.: (0484) 2377303, **Telefax:** 4051304, **E-mail: cochin@newagepublishers.com**
- **Guwahati** Hemsen Complex, Mohd. Shah Road, Paltan Bazar, Near Starline Hotel
Guwahati-781 008, **Tel.:** (0361) 2513881, **Telefax:** 2543669
E-mail: guwahati@newagepublishers.com
- **Hyderabad** 105, 1st Floor, Madhiray Kaveri Tower, 3-2-19, Azam Jahi Road, Near Kumar Theater
Nimboliadda Kachiguda, Hyderabad-500 027, **Tel.:** (040) 24652456, **Telefax:** 24652457
E-mail: hyderabad@newagepublishers.com
- **Kolkata** RDB Chambers (Formerly Lotus Cinema) 106A, 1st Floor, S N Banerjee Road
Kolkata-700 014, **Tel.:** (033) 22273773, **Telefax:** 22275247
E-mail: kolkata@newagepublishers.com
- **Lucknow** 16-A, Jopling Road, Lucknow-226 001, **Tel.:** (0522) 2209578, 4045297, **Telefax:** 2204098
E-mail: lucknow@newagepublishers.com
- **Mumbai** 142C, Victor House, Ground Floor, N.M. Joshi Marg, Lower Parel, Mumbai-400 013
Tel.: (022) 24927869, **Telefax:** 24915415, **E-mail: mumbai@newagepublishers.com**
- **New Delhi** 22, Golden House, Daryaganj, New Delhi-110 002, **Tel.:** (011) 23262368, 23262370
Telefax: 43551305, **E-mail: sales@newagepublishers.com**

ISBN: 978-81-224-0769-3
C-16-11-9781

Printed in India at Glorious Printers, Delhi.

NEW AGE INTERNATIONAL (P) LIMITED, PUBLISHERS
7/30 A, Daryaganj, New Delhi-110002
Visit us at **www.newagepublishers.com**
(CIN: U74899DL1966PTC004618)

Preface

This book is a part of 'Man and Environment Series' which focuses on various aspects of heavy metals in relation to our environment. In this volume, we have also described effects of some other elements which, by definition, could not be included with heavy metals but have similar toxic influences on the fauna and flora and, therefore, find a place with this view point.

Environmental pollution with toxic elements has attracted global attention. The threat to the quality environment associated with the constant dumping of non biodegradable metal/metalloid pollutants has drawn focus of scientists working in the trans-disciplinary areas of research. This volume covers the topics of heavy metals in biosphere, environmental pollution caused by them and their health effects.

It is intended to be used by students and scientists involved in the multi disciplinary areas of environment, biological sciences, medicine and pharmacology. It may also serve as a reference book for post graduate students of life sciences.

Authors wish to acknowledge with thanks the encouragement and facilities given by Hakeem Abdul Hameed, Prof. M. Amin, Ex-Vice Chancellor, Chancellor and Prof. Rasheeduzzafar, Vice Chancellor Jamia Hamdard, New Delhi.

Mohammad Athar
Shashi B. Vohora

CONTENTS

CHAPTER 1

INTRODUCTION

Man has made great strides towards industrial development and civilization which has without doubt improved the living conditions and comforts of human life. But, in the process, he has inadvertently upset the crucial environmental balance established by nature over a millenium. Human activity and consequent developments have brought about the spectre of an overwhelming degradation of all facets of the natural environment: physical, chemical, biological and social. This is adversely affecting the quality of life; paradoxically through the very means which are intended to improve it. Fortunately, awareness about environmental problems is growing worldwide and there is now global concern about the serious health risks involved and the need for preventive measures. Extensive efforts are being made for an indepth study of the environmental hazards and devise ways and means to resolve these problems. The scope of this book allows us only to discuss briefly the chemical aspects of the environmental pollution. During the process of evolution, organisms have developed their biochemistry in close connection with the composition of the external environment. However, both man and animals are incapable of adapting easily to any marked and quick changes in the chemical composition of their surroundings.

The concentration and distribution pattern of metals within the ecosystem is especially important. Metals and metalloids are ubiquitous. Of the 106 known elements, 82 belong to the group of metals, 6 to metalloids and 18 to non-metals. Metals are a unique class of toxicants since they cannot be broken down to non-toxic forms. Once the ecosystem is contaminated by them, they remain a potential threat for many years. Metals and metalloids occur in a variety of forms as ions, compounds and speciation complexes in the anthrosphere and are put to diverse uses by man (Table 1).

Rapid developments and increase in mining and industrial activities have gradually redistributed many of the toxic metals from the earth's crust to the environment. This has substantially raised the chances of human exposure to these metals (in excess of their natural levels) through ingestion, inhalation or skin contact. There is progressively rising evidence linking many diseases to the environmental factors which are injurious to biosystems. Most of these factors are man-made and among these human activity concerning heavy metals constitutes a major health hazard.

TABLE 1 INDUSTRIAL USES OF SOME HEAVY METALS/ METALLOIDS

METAL/METALLOID	Industrial Uses
ALUMINIUM	Buildings, automobiles, utensils, consumer goods, electric wires, packaging, fumigant.
ARSENIC	Pesticides / herbicides / rodenticides, glass and ceramic works, traditional medicines.
BERYLLIUM	Aircraft, space vehicles, rockets, radars, nuclear reactors, insulators for power transistors, beryllium bronze in electrical contacts, sparkless tools and bearing shells.
CADMIUM	Electroplating, storage batteries, stabilizers, solar cells, pigments, alloys for telegraph and telephone wires, photoelectrical and electron optical devices, nuclear reactors.
CHROMIUM	Alloying additive to stainless steel, chrome-plating dyes, pigments, tanning of leather, textiles, cassette tapes, magnetic storage media for computers, matches, pyro-technics, photography, seed disinfection.

COBALT	Glass, ceramics, paints, varnish, resins, alloys for tooth prosthesis, as cyancobalamine in medicine.
COPPER	Electrical goods, kitchenware, alloys, algicide/fungicide, pigments, supplement in deficient pastures, intra-uterine contraceptive devices.
GOLD	Coinage, jewellery, electronic industry, dentistry, medicine.
LEAD	Storage batteries, antiknock agent in petroleum/gasoline, paints, ammunition, glassware, ceramics, protection from radiation, bearing alloys, rubber industry, printing press.
MANGANESE	Alloys including steel, dry-cell batteries, diesel exhaust filters, fertilizers, pigments, pharmaceuticals, catalysts, dyes, paint, dryers, antiknock agent, disinfectants.
MERCURY	Thermometers, barometers, mercury vapour lamps, x-ray tubes, dental amalgam, chloralkali industry, electronic equipment, paints, military detonators, disinfectants, synthetic silk.
NICKEL	Alloys for corrosion-resistant equipment, cooking utensils, coinage, heating elements, gas turbines, jet engines, electroplating, paints, pigments, batteries.
PLATINUM	Jewellery, dentistry, scientific equipments, petroleum refining, industrial catalyst, cancer therapy.
SELENIUM	Glass industry, semiconductors, photocells, thermoelements, xerographic materials, rubber vulcanizing, lubricants, antidandruff agents, diagnostic scanning and labelling agent for malignant tumours, insecticide/fungicide, antioxidant in pharmaceutical industry.

SILVER	Photographic materials, coinage, jewellery, cutlery, mirrors, electroplating, disinfectant.
THALLIUM	Rectifiers, deep-temperature thermometers, low-melting glass with high refractive index, scintillation counters, superconductors, electric contacts, safety fuses, medical diagnostics, antidetonator.
THORIUM	Incandescent mantles, gas lights, alloys to coat tungsten wires and camera lenses.
TIN	Alloys, painting of steel, soldering, packaging, printer's alloy (type metal), dental amalgams, tooth pastes, fungicide, industrial biocide, PVC heat stabilizers.
URANIUM	Fission reactors, bomb fuel, armour piercing shells, ship blast and other military uses, electron microscopy, tile glazing.
VANADIUM	Tough, strong and heat- resistant steel, catalyst in sulphuric acid manufacture, chemical reagents, aniline dyes, coloured ceramics and glass.
ZINC	Galvanizing of iron and steel for corrosion protection, alloys, vulcanization of rubber, photocopying paper, paints, T.V. tubes, rayon glass, enamel and plastic industries, fertilizers, medicine and cosmetics.

The term heavy metal has variously been used to denote : (i) metals with atomic number 23 (i.e. vanadium) onwards except Rb, Sr, Y, Cs, Ba, and Fr; (ii) metals with density greater than 5; and (iii) metals which are toxic to man and other life forms when found in the environment. The eight most common heavy-metal pollutants listed by the Environment Protection Agency (EPA) are : As, Cd, Cr, Cu, Hg, Ni, Pb, and Zn. Since the subject of our concern is human health, we intend to discuss 23 elements in this

Fig. 1

PERIODIC TABLE SHOWING METALS / METALLOIDS OF MAJOR ENVIRONMENTAL CONCERN

RED LINE (   ) DIVIDES METALS FROM METALLOIDS AND NON-METALS.
MAGENTA LINE (   ) DEMARCATES HEAVY METALS (DENSITY >5)
BLACK NUMBERS AT TOP LEFT ARE ATOMIC NUMBERS. VALUES BELOW THE CHEMICAL SYMBOL REPRESENT DENSITIES. RED ASTRICES (*) DEPICT METALS / METALLOIDS OF MAJOR ENVIRONMENTAL CONCERN DEALT IN THIS VOLUME. ? : UN-NAMED

☐ Non-Metal
☐ Metalloid
☐ Metal

H																	2 He
3 Li	4 Be*											5 B	6 C	7 N	8 O	9 F	10 Ne
11 Na 0.53	12 Mg 1.74											13 Al* 2.70	14 Si	15 P	16 S	17 Cl	18 Ar
19 K 0.86	20 Ca 1.55	21 Sc 3.00	22 Ti 4.51	23 V* 6.10	24 Cr* 7.19	25 Mn* 7.19	26 Fe* 7.86	27 Co* 8.90	28 Ni* 8.90	29 Cu* 8.96	30 Zn* 7.14	31 Ga 5.91	32 Ge	33 As*	34 Se*	35 Br	36 Kr
37 Rb 1.53	38 Sr 2.60	39 Y 4.47	40 Zr 6.49	41 Nb 8.40	42 Mo 10.20	43 Tc 11.50	44 Ru 12.20	45 Rh 12.40	46 Pd 12.00	47 Ag* 10.50	48 Cd* 8.65	49 In 7.31	50 Sn* 7.30	51 Sb	52 Te	53 I	54 Xe
55 Cs 1.90	56 Ba 3.50	57 La 6.17	72 Hf 13.10	73 Ta 16.60	74 W 19.30	75 Re 21.00	76 Os 22.60	77 Ir 22.50	78 Pt* 21.40	79 Au* 19.30	80 Hg* 13.60	81 Tl* 11.85	82 Pb* 11.40	83 Bi* 9.80	84 Po 9.20	85 At	86 Rn
87 Fr	88 Ra 5.00	89 Ac															
			58 Ce 6.67	59 Pr 6.77	60 Nd 7.00	61 Pm	62 Sm 7.54	63 Eu 5.26	64 Gd 7.89	65 Tb 8.27	66 Dy 8.54	67 Ho 8.80	68 Er 9.05	69 Tm 9.33	70 Yb 6.98	71 Lu 9.84	
			90 Th* 11.70	91 Pa 15.40	92 U* 19.07	93 Np 19.00	94 Pu	95 Am	96 Cm	97 Bk	98 Cf	99 Es	100 Fm	101 Md	102 No	103 Lw	
			104 Ku	105 Ha	106 ?												

volume (Fig. 1). These include 21 metals (Ag, Al, Au, Be, Bi, Cd, Co, Cr, Cu, Fe, Hg, Mn, Ni, Pb, Pt, Sn, Th, Tl, U, V, and Zn) and two metalloids (As and Se). Two out of the 21 metals chosen (Al, Be) have densities less than 5 and are not heavy metals. The reason they are included in this book is their importance from the environmental health viewpoint and the chances of their exposure through occupation or environment or both.

CHAPTER 2

HEAVY METALS IN THE BIOSPHERE

2.1 DISTRIBUTION IN ECOSYSTEM

The biosphere is the natural environment of living objects. It envelops the Earth and contains surficial parts of the lithosphere, the lower part of the atmosphere, and the hydrosphere. A relatively homeostatic environment is essential for the survival of an organism in a given ecosystem. This necessitates the study of the chemical composition of air, water, and soil in order to monitor any abnormal changes due to industrial development and the consequent advancement of society.

2.1.1 Air

The ambient air that we breathe comes from the lower atmosphere (below 80 km). It is a mixture of nitrogen (78.08%) and oxygen (20.95%) containing other minor constituents such as argon (0.93%), carbon dioxide (0.03%) and some trace gases (e.g. neon, helium, methane, krypton, nitrous oxide, hydrogen, ozone, xenon etc.) with their levels ranging between 0.09 and 18 ppm. In addition, pollutants such as dust, smoke, soot, industrial and automobile exhaust and gaseous and particulate matter, may be found in varying concentrations depending upon factors, such as population, vehicular density, and location and type of industrial units installed in that area. Metal compounds, though not a normal constituent of air, are found in a great variety of forms particularly in urban areas. With some exceptions, they occur mostly in the particulate phase. The sources of metallic contaminations in air are both natural (e.g. terrestrial, marine, volcanic, biogenic) as well as anthropogenic (e.g. combustion, industrial, automobile). Man breathes about 12,000 litres of air each day, yet, unthinkingly, he keeps on using the sky as a refuse bin. As a result. the input of toxic trace metals in the atmosphere is increasing, thus causing progressive deterioration of air quality. Urban industrial air may contain suspended particulates of a wide variety of materials including heavy metals, such as lead, nickel, zinc, copper, cadmium, etc. depending on the occurrence of specific type of mining, milling or manufacturing units located in that area. Particles of 0.1 - 10 μm in diameter settle very slowly and remain suspended in the air for a longer duration of time. The typically small size (especially the particles < 1 μm in diameter) of these particulates allows them to be inhaled deeply into the respiratory system. Metal ions are absorbed by the lungs ten times more efficiently than the intestines. They are, therefore, of grave concern as a health hazard specially in densely populated areas and work places where high levels of metal-containing vapours and particles persist. The levels show an increasing trend in different areas as follows:

remote < rural < urban < industrial areas.

There is evidence to show that high emission of toxic trace-metals leads to an enrichment of the respective components even in remote environments such as high alpine or arctic areas.

The atmospheric occurence of metals depends on : (i) the source strength, (ii) atmospheric dispersion, and (iii) the deposition processes. The source strength, in turn, depends on the emission factors. Emission of trace metals in the northern hemisphere (80%) is much higher than in the southern hemisphere (30%). A sharp increase in anthropogenic emissions has been observed during the last two decades, with changing metal cycles and environmental perturbations on global (lead, vanadium, arsenic, tin), regional, i.e. upto 1000 km (zinc, cadmium, mercury, antimony, copper, silver, germanium, nickel, chromium) and local, i.e. upto 100 km, (cobalt, manganese, molybdenum, beryllium, uranium, titanium) scales.

Table 2 lists representative concentrations of some potentially hazardous metals/metalloids for vicinal air over Europe, North America and northern parts of India (Ludhiana, Chandigarh). The data represent atmospheric concentrations for urban, but not for highly industrialised regions. Values for atmospheric concentrations over Greenland are presented as baseline data for comparison.

A perusal of the data indicates a tremendous global increase in air concentrations of these potentially hazardous heavy metals in urban areas, as compared to a representative remote and unpolluted region, viz, Greenland. The World Health Organization (WHO) guidelines for air quality for some of these metals (concentration per m^3 , average time of exposure) are given as follows: cadmium (10-20 ng one year), lead (0.5-1 µm, one year), manganese (1 µg, one year), mercury (1 µg, one year) and vanadium (1 µg, 24 hours). The available data for air concentrations for India reveal consistently higher values, a matter that calls for remedial measures.

TABLE 2 ATMOSPHERIC CONCENTRATIONS (ng/m³) OF SOME METALS/METALLOIDS IN DIFFERENT PARTS OF THE WORLD

METAL/ METALLOID	GREENLAND	NORTH AMERICA	EUROPE	NORTH INDIA
Aluminium	NA	1	1	NA
Arsenic	0.007	2-40	1.5-53	NA
Beryllium	NA	0.2	2	NA
Cadmium	0.003-0.63	21	20	330-21,000
Chromium	0.6-0.8	60	25	11-194
Cobalt	NA	3	2	NA
Copper	0.03-0.06	100	340	208-2,682
Iron	166-171	829-2,000	130-5,900	NA
Lead	15-22	2,700	120	25-3, 677
Manganese	2.8-4.5	150	43	27-3,259
Mercury	NA	2	1	NA

Nickel	NA	100	25	12-321
Silver	NA	1	1	NA
Tin	NA	20	20	NA
Uranium	NA	0.2	0.02	NA
Vanadium	0.8-1.4	400	30	NA
Zinc	18-41	500	1,200	241-8,549

Compiled from: 1. Kabata Pendias, A. and Pendias, H.: *Trace Elements in Soils and Plants*, CRC Press, Boca Raton, Florida, 1984; 2. Sigel, H.: *Metal Ions in Biological Systems*, Vol. 20, Marcel Dekker Inc., N.Y. 1986; and 3. Kumar, R. : *Environmental Pollution and Health Hazards in India*, Ashish Publishing House, New Delhi, 1987.

Data represent median values or ranges.

NA: not available.

2.1.2 Water

Absolute pure water (which does not contain any soluble impurity) does not exist in nature. Water for drinking and domestic purposes is generally supplied by rivers, lakes, wells and springs. Such waters differ substantially from sea water and may contain organic matter from micro-organisms, salts of calcium, iron, magnesium, potassium, and sodium and traces of carbon-dioxide, oxygen, nitrogen, ammonia and other gases from the atmosphere. Presence of potentially hazardous heavy metals in water should be considered abnormal and this usually affects its appearance and palatability. The concentration of these metals in rain-water in urban areas may sometimes exceed safe levels for human consumption.

Metals of prime environmental concern e.g. cadmium, copper, lead, mercury, zinc, etc.) are present in very low concentrations in sea-water, but human influence is increasing these levels rapidly. Evidence of oceanic perturbation in the case of lead is unequivocal, with more than 20-fold difference between concentrations in polluted and unpolluted regions of the world. Fresh-water ecosystems are influenced by air-heavy-metal levels and additions from anthropogenic sources. The chemical forms of these metals in water are accessible to the biota through significant accumulation in the food chain. It is important to note that at lower concentration of dissolved metals in natural water, even smaller alterations in aquatic environment may critically influence the stability of the ecosystem. Besides the absolute concentration, the physico-chemical state and speciation in an aqueous system should also be looked into for any assessment of heavy metals in water for human use.

Table 3 gives data on the occurrence of some heavy metals in sea and fresh natural waters. A perusal of the data shows that the ratio of fresh-water to sea-water concentrations exceeds 100 for aluminium, lead and possibly, also for beryllium, indicating significant amount of these metals in fresh waters (rivers) relative to sea water.

TABLE 3 OCCURRENCE OF SOME HEAVY METALS IN SEA AND FRESH NATURAL WATERS

METAL/ METALLOID	CONCENTRATION ($mg\ m^{-3}$)	
	SEA WATER	FRESH WATER
Aluminium	2	300
Antimony	0.2	0.2
Beryllium	0.006	0.3
Cadmium	0.1	0.1
Chromium	0.3	1.0
Cobalt	0.02	0.2
Copper	0.3	3
Lead	0.03	3
Manganese	0.2	8
Mercury	0.03	0.1
Molybdenum	10	0.5
Nickel	0.6	0.5
Silver	0.04	0.3
Tin	0.004	0.009
Uranium	3	0.4
Vanadium	2.5	0.5
Zinc	5	15

Source : Bowen, H.J.M. : *Environmental Chemistry of the Elements*, Academic Press, London, 1979.

Drinking being one of the routes of intake of heavy metals into the human body, it has drawn special attention of health managers. The WHO has laid down the guideline values (mg/1) for heavy metals/metalloids in drinking water which is given as follows : arsenic (0.05), cadmium (0.005), chromium (0.05), lead (0.05), mercury (0.001) and selenium (0.01). Guidelines for radioactivity have also been specified : gross alpha activity : 0.1 Bq/1 (naturally-occurring alpha-emitting radio nuclides : ^{226}Ra, ^{224}Ra, ^{210}Pb, ^{210}Po, ^{232}Th, ^{234}U, ^{238}U; gross beta activity : 1 Bq/1 (beta emitting radio-nuclides having high toxicity : ^{90}Sr, ^{89}Sr, ^{137}Cs, and ^{60}Co). These guideline values represent levels which ensure aesthetically pleasing water* that does not result in any significant risk to the consumer's health. The crossing of these limits should be considered as a signal to investigate the cause of such an increase and take remedial measures. With a view to protect public health, water-quality standards have also been prescribed in India. The limits prescribed by the Ministry of Health, Government of India, for certain heavy metals in drinking water are given in Table 4.

TABLE 4 LIMITS PRESCRIBED FOR HEAVY-METAL CONTENT IN DRINKING WATER IN INDIA

METAL/METALLOID	CONCENTRATION (mg/1) PERMISSIBLE	EXCESSIVE
Arsenic	—	0.2
Chromium (hexavalent)	—	0.05
Copper	1.0	3.0
Iron	0.3	1.0
Lead	—	0.1
Manganese	0.1	0.5
Selenium	—	0.05
Zinc	5.0	15.0

2.1.3 Soil

Man depends on soils uncontaminated by toxic elements to produce his food. For terrestrial ecosystems, it is the point of entry of most materials into living matter. Soil is composed of finely divided rocks mixed with decayed

*Clear, sparkling, palatable, without any foul/stale odour or suspended particulate matter.

vegetable and animal matters constituting the portion of the earth in which plants grow. The earth's crust has an average thickness of 40 km in the continents and 7 km underneath the oceans. Though it represents only 0.4% of the total mass of our planet, agricultural productivity, and hence survival of mankind, is tied to this part of earth, thus, necessitating the study of its composition and functions. The elemental composition of soil is largely dependent on that of the rocks from which it is formed. Wide variations in the soil composition, therefore, occur in different parts of the world. Besides, it is affected by a number of factors, such as weathering, soil-forming processes, climatic conditions, geochemical affinities, sewage sludge, waste treatment, discharge from factories, fertilizers, pH, acid deposition, redox potential, amount and kind of organic matter, mining activity, aerial pollutants, etc. Though generalization has its limitations, Table 5 presents concentrations, calculated from global data, of some heavy metal/metalloids in uncontaminated surface soils.

Soil has complex functions which are beneficial to man and other living organisms. It acts as a filter, buffer, storage and transformation system and thus protects the global ecosystem against the adverse effects of environmental pollutions. These functions can be performed effectively only if the normal soil properties are preserved and the natural balance is not unduly disturbed.

TABLE 5 HEAVY METALS/METALLOIDS IN THE SURFACE OF SOIL OF THE WORLD

METAL/METALLOID	CONCENTRATION (mg Kg^{-1}) MEAN	RANGE
Aluminium	71,000	--
Arsenic	8	0.7-30
Beryllium	0.3	--
Cadmium	0.5	0.01-1.1
Chromium	60	20-100
Cobalt	8	0.1-29
Copper	23	1-80
Lead	30	9-84
Manganese	550	10-1500

Mercury	0.11	0.01-0.1
Nickel	20	5-35
Selenium	0.4	0.1-1.0
Silver	0.05	--
Tin	4	--
Uranium	2	--
Vanadium	87	10-460
Zinc	60	15-220

Source : 1, Bowen, H.J.M., *Environmental Chemistry of Elements*, Academic Press, London, 1979; 2. Kabata Pendias A. and Pendias, H. Trace *Elements in Soils and Plants*, CRC Press Inc., Boca Raton, Florida, 1984.

2.1.4 Plants

Plants absorb metals from soil, water and air. However, the chief source of metal absorption is soil. Uptake from it depends not only on the total content of the respective metal but also on its accessibility to roots and transfer across the soil-root interphase. The total amount of metal in a soil is, in turn, affected by the inherent natural resources of the particular area in addition to the agricultural and industrial activities.

Accessibility to roots depends on a variety of factors : (i) Soil : type, pH, drainage status, presence of organic matter, sewage sludge and microbial activity, (ii) Plant : species, part, season of collection, and (iii) Metal/metalloid: chemical form, location, etc. Metals present in the ionic state in the soil solution are readily available, while those which are firmly bound to rock minerals are the least available. Accessibility of Co, Mn and Ni increases with decrease in pH, while that of Mo and Se increases with rise in pH. Uptake of metals like Cu is only marginally affected by alterations in pH of the soil. Poor drainage enhances release of Co, Ni and V as also accumulation of organic matter which influences microbial activity and metal uptake.

Wide variations in metal uptake are observed in different plant species. Some plants are known to have special affinity for accumulating certain metals, e.g., *Astragalus* sp. (Se), *Crotalaria cobalticola* (Co), *Phaseolus vulgaris* roots (Zn), alga *Chlorculla vulgaris* (Au), *Alyssum* sp. and *Sebertia accuminata* (Ni), etc. The last mentioned plant is a tree native to Caledonia (site of world's largest Ni deposits). Its leaves and latex may contain respectively upto 10 g and 250 g per kg of Ni compounds on a dry-weight basis. Aquatic plants accumulate several hundred fold amounts of Ag as compared to other plants. Some Se-accumulating plants may attain a concentration of upto 1000 to 10,000 ppm of this metalloid. Young leaves and seed-heads reportedly have higher Cu and Zn concentrations while roots and fruit peel contain more Pb than other parts of the plant. The soil-root interphase is not a mechanical sieve but an active biological process tuned to the metabolic requirements of the plant. Chelation or precipitation near root may, however, affect metal uptake to some extent.

Besides the root, metal may enter the plant through aerial parts including the leaf surface. High Pb content (from motor exhaust fumes) may be observed in plants growing near busy traffic points. Minute particles (< 2 μm in diameter) of some metals/metalloids, e.g. As, Cd, Cu, Se and Zn, emanating from combustion sources, may get dissolved in rain water and enter the plant through leaf surface. Foliar application of some fertilizers may contribute to the bio-availability of Cu, Fe, Mn and Zn. Metals exhibit variable affinity to be absorbed by plants. The bio-accumulation index (plant/soil ratio) for various metals is reported to be as follows : Cd, Zn (1 to 10), Cu, Mo, Ni (0.1 to 1.0), As, Mn, Pb (0.01 to 0.1) and Fe, V, Ti (0.001 to 0.01).

From the foregoing discussion, it is clear that an interplay of a number of factors is involved in metal-uptake by the plants. This makes it virtually impossible to make general statements or present simplistic tables for the heavy-metal content of plants. Some metals/metalloids, however, are considered essential for the normal growth and metabolic functions of plants. These cannot be substituted by other elements in their specific biochemical roles and include Co, Mo (as enzymes cobalamide and nitrate reductase for nitrogen fixation), Cu, Zn (in oxidases and anhydrases for the metabolism of proteins and carbohydrates), Fe (involved in valence changes and photosynthesis), Mn (various enzymes with a role in O_2 photoproduction), Ni (as urease in hydrogenase metabolism) and V (as porphyrins in lipid metabolism).

Plants have a higher capacity to take up metals from soils or atmosphere than their physiological needs. While this may not have any adverse effect on plant itself, it may expose the consumer to higher intake of concerned metals. The critical concentrations of metals/metalloids in plant tissues are shown in Table 6.

TABLE 6 CRITICAL CONCENTRATIONS OF SOME METAL/ METALLOIDS IN PLANT TISSUES

METAL/METALLOID	CONCENTRATION (mg kg^{-1} dry matter) AT WHICH NO SENSITIVITY OR TOLERANCE DEVELOPS IN IMMATURE PLANTS
Arsenic	5-20
Cadmium	5-30
Chromium	5-30
Cobalt	15-50
Copper	20-100
Mercury	1-3
Nickel	10-100
Zinc	100-400

Sources : Kabata-Pendias, A. and Pendias, H. In : *New Horizons of Health Aspects of Elements* (eds. S.B. Vohora and Jan W. Dobrowolski), Ed. 1, p. 103, Jamia Hamdard, New Delhi, 1990.

Since plants are the major source of human food, a study of metal-content in edible plants should receive special attention. Table 7 provides this information about the presence of 16 metals/metalloids in some commonly used plant foods. Some varieties of wild mushrooms are reported to have high contents of As, Cd, Hg, Se, V (*Amanita muscaria* species may contain 100 times more V than other mushrooms or higher plants), and Zn. Tea, coffee and cocoa are rich sources of several metals including Al, Cd, Fe, Mn, Ni, and Zn. Some examples of rich sources of individual metals/metalloids are : radish (As, Cu), tobacco (As, Cd),

TABLE 7 METALS IN COMMON PLANT FOODS (mg/kg, fresh weight)

METAL/METALLOID	CEREALS	POTATOES	VEGETABLES	FRUITS	MUSHROOMS	TEA	COFFEE
Aluminium	1.2-5.4	1.2-2.5	0.1-5.0	0.4-2.8	4.0	128.0	30-45
Arsenic	0.05	0.07	0.03-0.04	0.02-0.04	0.14	0.13	0.22
Cadmium	0.016-0.75	0.03-0.05	0.04-0.067	0.008-0.011	0.46	0.21	0.017
Chromium	0.13-0.4	0.05-0.3	0.11	0.07	0.06-0.135	0.95	0.025
Cobalt	0.02-0.03	0.013	0.05-1.0	0.02-0.32	0.17	0.6-1.0	0.34-0.88
Copper	2.4-5.0	1.57	1.0 (Radish 8-22)	0.6	4.9	21.8	13.3
Iron	26-46	8.0	14.7	5.3	NA	172	168
Lead	0.035-0.18	0.09	0.07-0.2	0.06-0.12	0.3	0.73	0.20
Manganese	11-34	1.52-2.20	0.5-2.0	0.3-7.4	734	15.0	
Mercury	0.006-0.03	0.006	0.005-0.014	0.004-0.013	0.34	0.025	0.04
Molybdenum	2.55	NA	0.01-0.04	0.002-0.16	0.032	0.13	NA
Nickel	0.12-0.34	0.05-0.076	0.05	0.08	NA	5.25	0.77
Selenium	0.025-0.38	0.025	0.027	0.023	0.03-13.0	0.04	0.034
Thallium	0.02-0.08	0.025	0.005-0.13	0.05-0.14	0.048	NA	NA
Vanadium	0.08-2.3	0.01-0.02	0.02-1.0	0.008-0.06	0.001	NA	NA
Zinc	13-41	8.51	4.51	1.56	9.7-280	35.0	6.7

Values represent mean or range. Cereals : wheat, rye, rice, vegetables : lettuce, cabbage, tomatoes, spinach, celery, cucumber, Fruits : peaches, bananas, apples, citrus fruits.

NA : Not available.

Sources : 1, Weigert, P. In : *Metals and their Compounds in the Environment* (ed : E. Merian), Ed. 1, pp. 449-466, VCH, Wienheim, 1991 : 2. Ganapathy, S.N. In : *Proceedings of the First International Conference on Elements in Health and Disease* (eds : R.B. Arora, S.B. Vohora and M.S.Y. Khan)., Ed. 1, pp. 235-245, WHO and Institute of History of Medicine and Medical Research, New Delhi, 1984.

blue poppy seeds (Cd), unrefined beet and cane sugar (Cr), spinach (Fe), wheat (Zn), legumes and nuts (Fe, Mn, Se), soyabean, yeast (Se as selenomethionine), mustard (Al) etc. Arsenical insecticide sprays may contaminate grapes with high contents of arsenic. In addition, wines made from such grapes may have high levels of arsenic. Foliar sprays affect Hg levels. Utensils, cutlery and cans may add significant amounts of Al, Cu, Cr and Sn to plant foods.

2.1.5 Animals

Animals are exposed to metals via food (plants, other animals), drinking water and through inhalation. The mechanism of metal uptake is usually fast and efficient; the concentration factor (ratio of the metal concentration inside an organism to that outside it) is usually low. It depends on a number of factors which include the metal itself, its speciation, the organism and the environmental conditions. Though data for whole body metal/metalloid content in animals has its limitations in practical sense, some estimates are presented in Tables 8 and 9.

TABLE 8 METAL/METALLOIDS PRESENT IN WILD ANIMALS

METAL/METALLOIDS	WHOLE BODY CONTENT (ppm)
Arsenic	0.4-1.1
Cadmium	0.1-0.3
Cobalt	0.5-0.2
Chromium	0.11-0.48
Copper	1-9
Iron	60
Manganese	0.7-2.4
Nickel	0-0.2
Lead	0.25-0.8
Tin	0-2.2
Tellurium	0-0.4
Titanium	0.3-3
Vanadium	0-2
Zinc	22-30

Source : Schroeder, H.A., *The Trace Elements and Nutrition : Some Positive and Negative Aspects*, pp. 19-37, Faber and Faber, London, 1976.

TABLE 9 WHOLE BODY CONTENT OF COPPER, IRON AND ZINC IN SOME LABORATORY AND DOMESTIC ANIMALS

METAL	CONCENTRATION (mg/kg fat free tissue)									
	MOUSE		RAT		RABBIT		CAT		PIG	
	Adult	New Born	Adult	New Born	Adult	New Born	Adult	New Born	Adult	New Born
Cu	8.9	6.7	2.0	4.3	1.5	4.0	1.5	2.9	2.5	3.2
Fe	90	67	60	59	60	135	60	55	90	29
Zn	50	46.4	30	22.5	50	22.5	23	28.7	25	10.1

Source : Widdowson, E.M. and Dickerson, J.W.T. In : *Mineral Metabolism : an Advance Treatise* (eds. C.L. Comer and P. Bronner), vol. 2, Part A, Academic Press, New York, 1964.

The essential and toxic roles of metals have been studied extensively in recent years. The following metals/ metalloids are considered essential for animals : As (non-specific growth stimulation), Co (constituent of vitamin B_{12}), Cr (regulator of metabolism of glucose and cholesterol), Cu (constituent of oxidases which are important for the regulation of redox reactions, respiration, cartilage formation), Fe (in hemoglobin, cytochromes, catalase, peroxidases), Mn (as arginase, superoxide dismutase and pyruvate carboxylase in urea synthesis, protecting mitochondria from free radical damage and citric-acid cycle), Mo (aldehyde and xanthine oxidases in formation of fatty acids and uric acid respectively), Ni (constituent of several enzymes, undefined growth stimulation), Se (as glutathione peroxidase, role in tissue oxidation and cardiomyopathy), Sn (as a constituent of gastrin with digestive and growth-promoting functions), V (in lipid metabolism and bone mineralization) and Zn (constituent of several enzymes : proteases, anhydrases, superoxide dismutase with role in protein biosynthesis, energy metabolism, protection against damage by superoxide radicals, fertility, etc.). Essentiality for Ba (deficiency has been reported to cause reduced calcification of bones and teeth and growth retardation in rats and guinea pigs), Cd (growth stimulatory effect through increased activity of elongation factors 1 and 2) and Ti (inhibits hydrolytic enzymes and as titanate can elicit growth stimulatory effects) needs confirmation.

The difference between deficient and toxic levels for most metals generally falls within a narrow range of about two orders of magnitude. The recorded ranges of deficient and toxic values sometimes overlap because, under the influence of various external and internal factors, a concentration considered deficient in one situation may become toxic as the conditions change. Mechanisms, such as chelation and formation of protein complexes, are found to transform toxic ionic metals to non-toxic bound forms suitable for uptake, transport and storage. Plants may sometimes partly protect man and animals from the adverse effects of metals ingested as food. The comparative picture for metal toxicity in plants and mammals is represented as below :

Toxicity to plants :

Hg > Ni > Pb > Cu > Cd > Cr > Sn

Toxicity to mammals :

Hg, Cd > Be > Pb > Cu > Se > Ni, Cr > Zn, V

As animal products are extensively used as human food, they must be examined for any traces of hazardous metals (Table 10).

2.1.6 Micro-Organisms / Biomethylation

Metal uptake by micro-organisms may occur through the processes involving (i) non-specific absorption and/or (ii) metabolically mediated mechanisms. While the latter processes are observed in viable cells only, non-specific binding on the cell surface may occur in both viable and non-viable cells. Metal cations are readily absorbed to anionic ligands (carboxyl, hydroxyl, phosphoryl and sulfhydryl groups) at the cell surface through a rapid, reversible process, which is usually independent of temperature/energy metabolism. Metabolically mediated mechanisms chiefly involve divalent and, to a lesser extent, monovalent cation transport systems. Some metals, such

TABLE 10 METAL/METALLOID CONTENT IN FOOD ITEMS OF ANIMAL ORIGIN

METAL/ METALLOID	CONCENTRATION (per kg fresh weight) MILK AND DAIRY PRODUCTS	MEAT, PORK, BEEF, CHICKEN	FISH, OYSTERS, SEA FOOD
(1)	(2)	(3)	(4)
Aluminium	Milk : 0.8-2 mg/1	Meat and liver : 5 mg Cooked beef : 0.2 mg	Cooked cod fish: 0.4 mg
Antimony	Milk : <0.1 μg/1	Higher concentrations occur in liver, particularly from old animals : 2 mg	Oysters: <7 μg
Arsenic	Milk : < 50 μg/1	0.01 to 6.3 mg. In terrestrial species, it occurs in the form of dimethylarsinic acid, methyl arsonic acid and inorganic As (III) and As (V) compounds. Liver and kidney : 0.05-15 mg but concentrations upto 10-15 mg may be observed after exposure.	High content has been reported in sea food where it occurs in stable non-toxic forms: arsenobetaine, arsenocholine, arsenolecithin etc. Marine fish : 1-10 mg Certain bottom feeding fish and crustaceans : > 100 mg Prawns : upto 175 mg Mussels : upto 120 mg Oysters : upto 3.7 mg
Beryllium	Milk : < 0.75 μg	Mammalian muscles : 0.75 μg Mammalian bones : 3 μg	Salt water fish: 2-200 μg
Bismuth	Milk : 2.2 μg/1	Land animals : 4 μg Bovine liver, contains higher amounts.	Marine animals : 40-300 μg
Cadmium	Highly variable Milk: 0.21-37 mg/l	Meat & chicken: < 5 μg Kidney & liver: < 200 mg	High content in Oysters.
Chromium	Milk, milk Products : 10-60 μg	Meat, Pork, Beef Chicken : 0.1-30 mg Eggs : 1.91 mg Mammalian : bones: 0.1-30mg Liver and kidney : 0.2-0.4 mg	Sea Fish 0.03-2 mg Sea foods: 0.05-0.07 mg Oysters have higher content.

(1)	(2)	(3)	(4)
Copper	0.19-3.92 mg	Meat, Pork, beef, Chicken : 1.19-7.13 mg Eggs : 2.44 mg Liver : 20-190 mg (mean : 50 mg).	Fish : 0.7-15 mg Oysters : upto 137 mg Other sea foods : 3.33-3.4 mg
Iron (per 100 kcal)	0.7-1.9 mg (300-400 mg/1)	High conc. in organ meats and egg yolk.	Shell fish : > 2 mg
Lead	Milk:2-90 μg/1	More in organ meats like liver and kidney.	Fish : 0.01-0.03 mg, upto 0.5 mg in some fishes. Fish liver: 0.04-0.15 mg Oysters : 1 mg
Manganese	0.01-1.8 mg (mean : 0.7 mg)	Mammalian muscles : 0.2-0.3 mg Liver : 10 mg	Sea Fish : 0.3-4.6 mg Sea foods: 0.12-0.72 mg (without Oysters)
Molybdenum	Generally low Mo content is found in animal products.		
		Meat : 20-50 μg Eggs : 90 μg	
	50-110 μg (mean : 70 μg)	Liver contains comparatively higher amounts : 1-3 mg Kidney : 0.3 mg	Sea foods : 20-100 μg
Mercury	Milk : < 1-19 μg/1	Muscles of land animals rarely contain > 50 μg Beef : 10-20 μg Chicken : 40 μg Eggs : 30 μg Liver : upto 100 μg	Figures in parenthesis indicate methyl mercury concentrations : Mussels 20 μg (8 μg) Shell fish 20 μg (18 μg), Open-ocean predators, e.g. sword and tuna fish : (1 mg and upto 10 mg if the water is industrially contaminated).

(1)	(2)	(3)	(4)
Nickel	Milk : 10-30 mg/1	Meat : 1-13 μg Bovine liver : 70-180 μg	NA
Selenium (per kg dry weight)	10-100 g	Beef : 0.35 mg Pork : 0.52 mg Chicken : 0.21 mg Eggs : 1.8 mg Liver and kidneys : 1.57-5.97 mg	Fish : 1.4-2.9 mg
Silver (per kg dry weight)	Milk : 25-54 mg/1	Mutton, Beef, Pork : 4-24 μg	11 mg (fresh weight)
Thallium	NA	Pork: 0.028-0.138 mg Kidney: 0.063-0.244 mg Eggs: < 0.1-0.7 mg	NA
Tin	Milk : 0.3 μg/1	NA	NA
Vanadium	Milk:0.4 μg	Liver : 2.4-4.4 μg Pork and Veal : 0.1 μg in ash	Salt water fish : 30 μg Sea foods : 2.4-4.4 μg
Zinc	Milk : 3-5 mg	Meat and meat products : 20-60 mg Liver : 100-150 mg Kidney : 50-100 mg	Fish : 15 mg Oysters : 100-200 mg (upto 2 g in some samples).
Zirconium	NA	Mammalian muscle : 80 μg	NA

NA : Not available.

Sources : 1. Iyenger, G.V. *Elemental Composition of Human and Animal Milk*, IAEA/TECDOC 269, International Atomic Energy Agency, Vienna, 1982; 2. Ganapathy, S.N. In : *Proceedings of the First International Conference on Elements in Health and Diseases* (eds : R.B. Arora, S.B. Vohora and M.S.Y. Khan), Ed. 1, pp. 235-245, WHO and Institute of History of Medicine and Medical Research, New Delhi. 3. Merian, E. *Metals and their Compounds in the Environment*, p. 1056, VCH, Weinheim, 1991.

as Cd, Co, Mn, Ni, and Zn can replace Mg in several enzymes/co-enzymes which play a role in phosphate transfer in several bacteria and fungi. Certain organisms (e.g., *Staphylococcus aureus*) utilize metabolic processes to exclude metal ions from the cell and establish resistance. Cadmium exclusion from the cell and resistance to *S. aureus* is plasmid mediated. Organisms which produce dihydrogen sulphide (H_2S) from insoluble sulphides on binding with a particular metal develop tolerance to it. This may also protect other organisms (non H_2S-producing) present in the surrounding environment. Alterations of the form of occurrence of metals by micro-organisms through chelation, complexion, methylation etc. significantly affect their bioavailability and toxicity. The presence of glutathione, citric acid, oxalic acid, humic acid, methionine and other amino acids may also reduce metal toxicity. An interplay of various factors, such as pH, temperature, redox potential, sunlight, composition of water and soil, number and type of organisms, period of exposure, interaction with other metals, etc., is responsible for marked differences between the toxicity of metals to micro-organisms in pure culture and in their natural habitat. The toxicity of metal ions on heterophilic bacterial populations, in general, reportedly exhibits the following order :

$$Ag^{+} > Cu^{2+},\ Ni^{2+} > Ba^{2+},\ Cr^{2+},\ Hg^{2+} > Zn^{2+},\ Cd^{2+}$$

Heavy metal and antibiotic resistance may co-occur due to common environmental factors. For example, pencillinase plasmids of *S. aureas* may also exhibit resistance to As, Bi, Cd, Hg and Pb. Since micro-organisms are generally first to be affected by discharges of heavy metals into the environment, they may serve as useful bioindicators of environmental pollution. For this reason, and due to their role in the accumulation of methyl mercury in the food chain, mutagenicity and carcinogenicity, the study of inter-relationship of micro-organisms and heavy metals has significant implications for human health and well being.

Biomethylation

The knowledge of biomethylation can be traced back to early nineteenth century. As early as 1824, the side effect called bismuth breath following the treatment of stomach ailments by bismuth carbonate were recognized. These were attributed to the presence of tellurium impurities in bismuth carbonate. This was presumably the result of biomethylation of tellurium to their dimethyl derivative. It was also observed that paints/pigments containing arsenic undergo biomethylation by modules to produce a gas with garlic-like odour. However, the precise nature of the methylated products could not be ascertained. Recently developed analytical techniques defined and quantitated exact nature of these molecular species. It is now known that a number of metals and metalloids may undergo biomethylation leading to their detoxication. On the other hand, many metals or metalloids, instead of being detoxified, become more hazardous to living beings including humans through this process.

Micro-organisms play an important role in the environmental cycles of some elements. Organometallic compounds in our environment are often produced by biomethylation where the transfer of a methyl group(s) is mediated biologically from a donor molecule to the inorganic form of the element. Microbes possess efficient mechanisms to carry out such relations. For example, various species of bacteria, fungi and algae are known to biomethylate a number of metals and metalloids, such as tin, lead, cadmium, mercury, thallium, arsenic, germanium, selenium, tellurium etc. Thus, the methylated forms of arsenic and antimony have been observed in natural waters while methyl mercury compounds are found in the body organs of a number of marine organisms. Bacteria and fungi in sewage sludge may methylate a number of above mentioned elements contaminating the residential surroundings.

In the human body, vitamin B_{12} is perhaps the most important entity which may perform biomethylation of metals/metalloids. However, the major portion of the body-burden is contributed as a result of exposure to the environment or by eating contaminated food, specially marine food products. Some examples of biological methylating organisms with their specificity or preference to methylate metals/metalloids are given in Table 11.

TABLE 11 METALS/METALLOIDS AND THEIR BIOLOGICAL METHYLATING ORGANISMS

METAL/METALLOID	NAME OF ORGANISMS
Arsenic	Candida lumicola, Methanobacter
Cadmium	Pseudomonas
Lead	Pseudomonas, Acinctobacter, Flavobacterium, Aeromonas
Mercury	Escherichia coli
Selenium	Candida lumicola, Pseudomonas
Tin	Pseudomonas

2.2 DISTRIBUTION IN THE HUMAN BODY

2.2.1 Whole Body

The bulk of the human body (99.9%) is composed of 12 elements (O : 65%, C : 18%, H : 10%, N : 3%, Ca : 2%, P : 1%, K, S, Cl, Na, Mg, and Fe : 0.9%). All other elements together constitute the remaining 0.1% (1 g/kg) of the body. With the exception of Fe, all other metals/metalloids of environmental concern fall into the latter category and may be classed as trace (Al, Cu, Fe, Hg, Zn, Zr) and ultra/micro trace (Ag, As, Au, Be, Bi, Cd, Co, Cr, Mn, Mo, Ni, Pb, Sb, U, V) elements. The total amounts present in an average adult human of 70 kg body-weight are shown in parentheses : Fe (4.2 g), Zn (2.33 g), Zr (300 mg), Al, Pb, Sb, (50-100 mg), As, Cd, Cu, Hg, Mn, Sn, V (10-50 mg), Au, Co, Cr, Mo, Ni, Se (1-10 mg), Ag, Bi (0.1-1 mg) and Be, U (<0.1 mg). Molar values for these quantities give a better picture and indicate that the body contains at least 10^{19} ions of each of these trace metals, with 10^5 to 10^6 ions in each cell assuming they are evenly distributed. However, as this assumption is not true, metabolically active cells may have much larger quantities of these metals

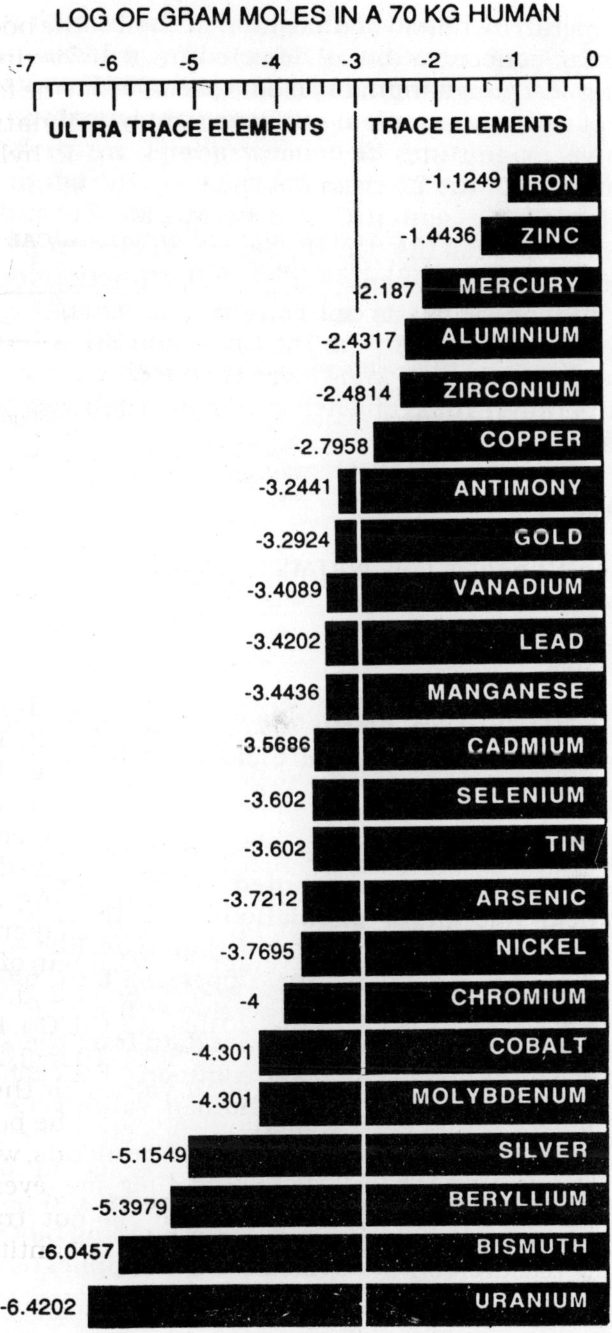

Fig. 2. Metal/Metalloid Content of Adult Human Body

The metal/metalloid composition of the human body using molar concentration is depicted on a logarithmic scale in Fig. 2. Though the data presented has its limitations because of wide variations in published analytical values, the relative magnitudes of concentrations are probably representative. Of the 23 elements shown in the figure, six may be classed as essential (Co, Cu, Fe, Mn, Mo, Zn) and six as possibly essential (As, Cr, Ni, Se, Sn, V) elements. The evidence for the requirements and essential functions of the latter in human beings is not certain. The remaining 11 metals (Ag, Al, Au, Be, Bi, Cd, Hg, Sb, U and Zr), as far as present state-of-the-art knowledge is concerned, have no known functions in man and hence are called non-essential elements.

2.2.2 Blood

Table 12 shows concentrations (range of values and means) of some metals/metalloids in the human blood.

2.2.3 Tissues

The distribution of metals/metalloids of major environmental concern in human brain, heart, liver, kidneys, lungs and skin is presented in Table 13.

A perusal of data in Tables 12 and 13 revealed wide variations which may be attributed to differences in the geographical habitat of populations and to analytical procedures. For example, some investigators consider serum Cr levels of 0.5 mg/ml as the upper limit in normal individuals, while others accept a value of 5 mg/ml as a definitive proof of Cr deficiency. Even data from the same laboratory may be inconsistent and confusing. For example, serum Al concentrations for 17 human volunteers was reported to be 800 ng/ml in 1962. Nine years later, the same laboratory reported these values to be 109 ng/ml for 68 subjects. Such variations in base-line values and lack of precision in analytical techniques are stumbling blocks in furthering research on trace metals and its applications to human welfare.

TABLE 12 METALS/METALLOIDS IN HUMAN BLOOD

METAL/METALLOID	CONCENTRATION (μg/1)			
	WHOLE BLOOD	RBC	PLASMA	SERUM
Aluminium	100-720 (390)	64-4600	280-410	110-780 (280)
Antimony	1.2-4.7	4.6	3.2-54.0	2.5-3.2
Arsenic	2.5-190	2.7	2.4-10.0	20-43
Beryllium	<0.1-3.8	NA	<4.0	NA
Bismuth	<2 - 9	NA	<0.6	<10
Cadinium	1.1-7.4 (5.2)	NA	2.6	2.3-12.0
Chromium	6.5-107	19-66 (37)	0.7-12.0	2-20
Cobalt	0.30-99	0.16-86	26-164	0.20-14.0
Copper	640-1280 (1010)	750-1310 (980)	610-1410 (1120)	970-1640 (1190)
Gold	42-420	0.8	0.06	0.08
Iron	301-530 mg/l (447)	985-1140 mg/ll	710-1270 (1100)	87-1870 (1090)
Lead	88-400 (214)	150-610 (460)	12-180 (130)	16-130
Manganese	16-75	11.5-66.0	0.59-68.0	0.54-61.0
Mercury	<5-20	7-50	2-11	12
Molybdenum	0.95-75	17.5	13.0	6-27
Nickel	4.8-106	48-106 (83)	10-66 (54)	2.6-40.0
Platinum	<40	NA	<40	<100
Selenium	57-320 (171)	71-340	<30-590	98-327 (122)
Silver	3.4-120	4.3-15.0	3.6 14.0	2.9-200
Thallium	0.47-9.0	NA	<2.5	<30
Thorium	<0.2-2.4	0.16	<0.04	<40
Tin	8.5-290	250-510	<4-105	30
Uranium	0.10-0.55	0.24	< 0.13-59.0	20
Vanadium	7.9-58 (13.6)	1.7-8.9	< 10-56	< 10
Zinc	4.8-9.3 mg/l (7.0)	7.6-16.1 mg/l (14.2)	790-1700 (1140)	670-1830 (1150)

Source : Iyenger, G.V., Kollmer, W.E., and Bowen, H.J.M., *The Elemental Composition of Human Tissues and Body Fluids*, Verlag Chemie, New York, 1978.

TABLE 13 METALS/METALLOIDS IN HUMAN TISSUES

Values represent range and/or means in μg/g for ash except when specified otherwise; d : dry, f : fresh; NA = not available

METAL/METALLOID	BRAIN	HEART	LIVER	KIDNEYS	LUNGS	SKIN
Aluminium	13-360	30-310	65-500	35-560	2500-4500 (3214)	400-1000
Arsenic	0.012d	0.021-0.027d	0.03 0.35d	0.007-0.05d	0.010-0.113f	0.092-0.124d
Beryllium	NA	NA	NA	NA	NA	0.02f
Bismuth	< 2	< 2	< 2	< 2	< 2	< 2
Cadmium	< 50	< 50	50-480 (203)	1300-5100 (2756)	< 50-80	< 50
Chromium	0.2-3.5	1.5-5.6	0.7-12.7	1-24	17-32 (23.3)	41
Cobalt	< 2	< 2-17	< 2-13	< 2	< 2-8	< 2-300
Copper	340-460 (401)	240-410 (350)	882	379	130-420 (220)	143
Gold	< 10	< 10	< 10	< 10	< 10	< 10
Lead	5-14	0.075-0.40f	94-150 (122)	36-140	28-82 (55)	17-46
Manganese	15-34 (22)	21-47 (27)	130-150 (130)	40-91 (79)	20-51 (29)	0.003-1.08d
Mercury	0.10-0.28f	0.005-0.15f	0.005-0.25f	0.063-2.75f	0.01-0.25f (0.20)	0.003-0.18f
Nickel	< 5	< 5	0.032-4.6d	< 5-12	< 5-24	47
Platinum	NA	0.0007f	NA	NA	NA	NA
Selenium	0.27-1.64f	0.18-1.14f	1.73-2.40d	2.05-3.5d	0.087-0.23f (0.15)	0.06-0.27f
Silver	< 0.5-1.70	< 0.1	< 0.10-1.70	< 0.1-0.5	< 0.1-0.5	1.3
Thallium	< 0.001f	< 0.1f	0.47d	< 0.003f	0.0013f	0.0006-0.03f
Throium	NA	NA	NA	NA	0.01f	NA
Tin	< 5	NA	< 5-23	< 5-16	0.17-1.75f	21
Uranium	0.00032-0.0016f	NA	0.0008f	0.002-0.005f	NA	NA
Vanadium	< 1	< 1	< 1	< 1	3-13	< 1
Zinc	800-1200 (915)	1635-3400 (2772)	5543	2873-8100 (5018)	990-2200 (1470)	615-1000

Source : Iyenger G.V., Kollmer, W.E. and Bowen, H.J.M. *The Elemental Composition of Human Tissues and Body Fluids*, Verlag Chemie, New York, 1978.

CHAPTER 3

ENVIRONMENTAL POLLUTION

3.1 SOURCES

Environmental pollution with metals perhaps began with the discovery of fire and gradually aggravated to its present alarming level with industrial development and advancement of society. Modern societies use many goods and amenities involving an increasing range of metallic products. High consumption, frequent disposal and replacement of disposable items are generating diverse types of metallic wastes. These wastes are invariably discharged into the environment and thus are poisoning the biosphere. Human activities, such as mining, smelting, refining, energy production, industrial and vehicular emissions, agricultural operations, sewage discharge and disposal of wastes are rapidly increasing environmental pollution. Prof. Jerome Nriagu, a Canadian environment scientist working at the National Water Research Institute in Burlington, conducted global surveys to assess the magnitude of such problems (Fig. 3.1). Between 1900 and 1980, a manifold increase in the mine production of many metals [examples : aluminium (114-fold), nickel (35-fold), chromium (18-fold), copper (5-fold) and zinc (4-fold)] was recorded. Industrial emissions of cadmium, copper, lead, nickel and zinc in the air registered about 8-, 6-, 9-, 51-, and 8-fold increases respectively during the same period. Eventually, all metals that are mined will be dissipated into the environment, thereby endangering human health. As metals cannot be degraded further to non-toxic products or recovered economically from such a contaminated environment using current technology, their deleterious effects tend to be permanent.

The environment touches us everywhere : the air we

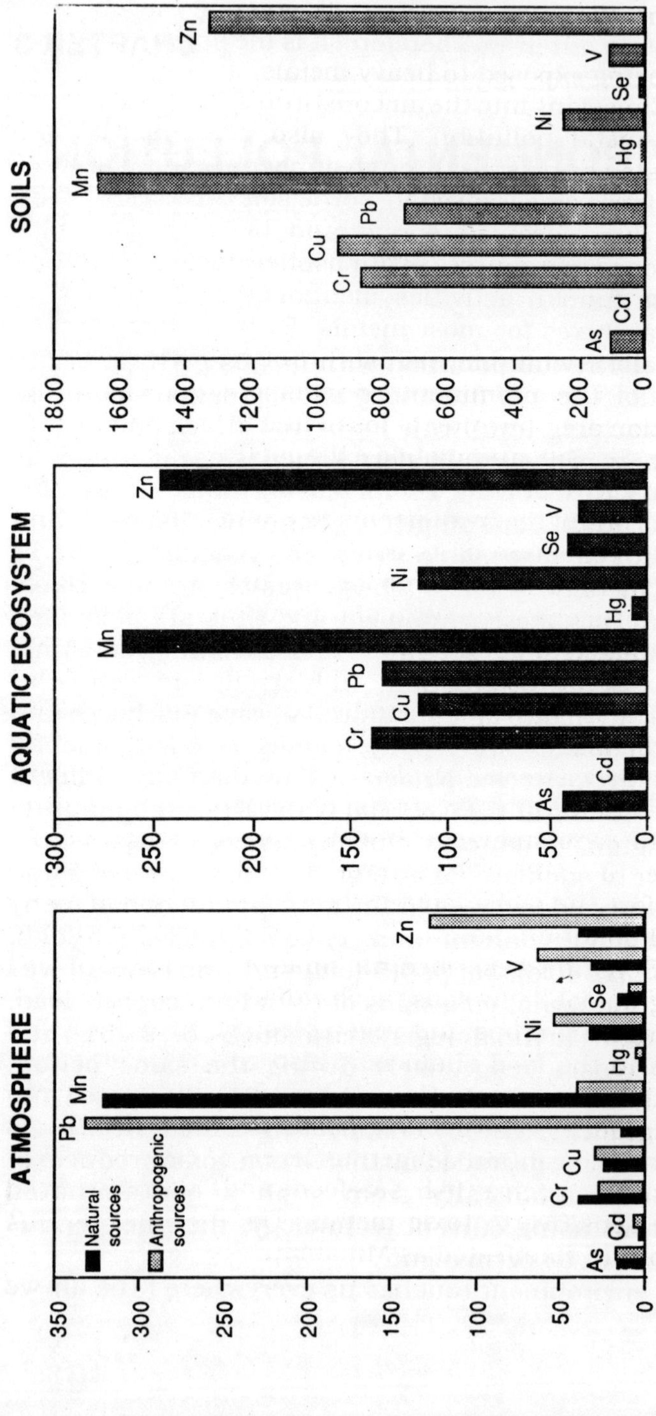

Source : Nriagu, J.O. Global metal pollution : poisoning the biosphere ? *Environment*, 32 (7), 7-11, 28-33, 1990.

Fig. 3.1. Global Input of Metals/Metalloids into the Environment

breathe, the soil we live on, the water and the food we take to sustain our lives. Therefore, it is inevitable to keep away from being exposed to heavy metals.

Emissions into the air constitute the greatest source of heavy-metal pollution. They also contaminate aquatic ecosystems and soils through atmospheric fallout. Natural emissions come from wind-borne soil particles, volcanoes, forest fires, sea-salt sprays and biogenic sources. The anthropogenic sources of atmospheric emissions, through diverse human activities mentioned earlier, exceed the natural fluxes for most metals. Even the metals emitted naturally in wind-blown dusts are often of industrial origin. Some of the prominent sources of atmospheric metal pollution are : burning of fossil fuel to generate energy (V, Ni, Hg, Se, Sn), automobile exhaust (Pb), insecticides (As), manufacture of steel (Mn, Cr), smelting (As, Cu, Zn), etc. Lead is one of the most pervasive environmental pollutant. Despite reduction of its content in petrol/diesel by many countries, it still accounts for roughly two-thirds of global lead emissions. One third of the world's urban population is estimated to be exposed to marginal or unacceptable air lead concentrations.

The sources of metallic contamination of aquatic ecosystems include sewage and industrial discharges, domestic waste waters, base-metal mining and smelting, electric power plants, atmospheric deposition, etc. Among aquatic ecosystems, fresh-water resources are particularly prone to external trace-metal inputs. These should, therefore, be considered to be at greatest risk from toxic metal pollution.

Soils are described as sinks for metals. The latter, being immobile in soils, accumulate in the top soil, thus endangering crops and vegetables. Besides plants, other items in the food chain may also get contaminated. The health authorities in Canada have advised people against eating kidneys and livers of moose, deer and bear. These are reported to contain dangerous levels of cadmium. Aquatic organisms, such as fish, sea-foods and fish-eating birds are quite sensitive to toxic metals. The example of mercury poisoning manifested as Minamata disease is well known.

Similar cases have also been reported among fishing villagers in other parts of the world.

Sources or factors which may cause or enhance heavy-metal pollution include acid rains, food grown in metal-contaminated areas, storing and cooking utensils and inadequate canning and preservation practices. Links between contamination of milk and drinking water by copper due to boiling/storage in brass/copper utensils and Indian Childhood Cirrhosis (ICC) have been reported by many investigators. Tin coating of brass utensils prevents Cu contaminations. Foods stored or cooked in aluminium pans, trays or foil accumulate some aluminium. The tin content of canned foods varies greatly; foods packed in lacquered cans contain markedly less Sn (4 μg/g) than those not coated with lacquer (40-150 μg/g). Lead pollution of drinking water may occur from freshly painted pipes.

Owing to economic pressures and industrialization, the threat of metal pollution is slowly becoming a reality in developing countries including India, thus necessitating the monitoring of exposure levels and remedial measures before it is too late.

3.2 BIOINDICATORS

Long-term monitoring of an ecosystem is necessary for assessing and improving the quality of air and water with the various stages of development. This helps in planning and controlling the effects of anthropogenic activities. Environmental pollution embodies much more than just raised levels of a pollutant. Important questions which should be kept in mind include, (i) whether the disturbance is acceptable or unacceptable, (ii) defining the levels of exposure which constitute an unacceptable disturbance, and (iii) how seriously does such an exposure affect the well-being of man. Biological materials react ecotoxicologically to contaminants in the form of measurable responses. The latter constitute a convenient substitute for man as a target organism. Biological materials such as bacteria, fungi, algae, plants, animals, marine organisms, cell cultures, placenta, foetus, blood, urine, hair, nails, etc. are used for the recognition and monitoring of responses to anthropogenic influences on the environment, and are termed bioindicators. In recent years, they have been widely used to monitor the emission and ill-effects of pollutants.

These indicators have been variously classified as : (i) optical, chemical, physical or biochemical; (ii) sensitive or accumulative; and (iii) local or universal. Optical indicators involve visual observations, such as reduced plant growth, decolouration of leaves, depressed food consumption and behavioural changes in animals. Chemical indicators measure the accumulation of the pollutants, such as heavy metals. Physical-biochemical indicators may inhibit enzyme activity or lead to derangement of physiological functions. Sensitive indicators are used in ecological surveys and hazard evaluation. They provide information on lethal or sub-lethal effects in most sensitive organisms or species, critical concentrations in cells/tissues/organs, conditions influencing toxicity, etc. Fishes have been often used to test the toxic effects of effluents introduced into aquatic ecosystems. Many plants and organisms amass polluting agents without affecting their chemical nature. Heavy metals (particularly Cd, Cu, Hg, Pb and Zn) have been extensively studied for bioaccumulation in many biological systems.

An ideal bio-monitor must have the following characteristics : (i) abundance in the area under study (both in polluted and non-polluted parts), (ii) capacity of retaining the pollutant progressively during the period of exposure, (iii) eliciting reproducible effects; (iv) should be easy to collect throughout the year, and (v) should be cheap. The indicator species (phyto- and zoo-indicators) may be universal or local. Comprehensive lists of plant indicators of mineral deposits and heavy-metal tolerant plants have been documented in literature. Metals tolerated by plants at concentrations dangerous to man and animals pose a significant health risk. Heavy-metal analyses of bird feathers have been frequently used for monitoring pollution in terrestrial ecosystems. Likewise, the analysis of dog hairs for Cd, Pb, and Zn content has been described as a zoo-indicator for determining spatial differences in mean amount of the metals.

Mammalian and human tissue cultures are not only important tools for evaluating biological, including cytotoxic, carcinogenic and immunosuppressive effects of pollutants, but they also provide a justification for a detailed *in vivo* investigation. Rodent macrophages have been used for assessing such effects. Human exposure may be directly studied by observing gross symptoms such as morbidity, mortality and by analysis of urine, blood, hair, nails, body fluids, erythrocytes, etc. for metal content. Of these, human serum is the most extensively used and accurate bioindicator. For example, a good correlation between air and blood lead contents has been observed. It is reported that about 1 $\mu g/m^3$ of Pb in air may contribute to 2 $\mu g/dl$ of blood Pb. Quantitative analyses of hair and nails may be very convenient to survey exposure levels in exposed population groups. The values, so obtained, parallel blood/serum concentrations for many metals. Elevated hair lead concentration is a good index for learning disability in children. Hairs, however, are often exogenously contaminated; metals may become strongly bound to SH-groups, thereby resisting complete removal by washing. Living beings are particularly sensitive to heavy metals during stages of differentiation and organogenesis. This makes the foetus a good indicator for most studies on neurotoxicity of mercury compounds. The most feasible biological specimens to be sampled for the assessment of exposure to individual metals are : blood, serum or plasma (Cd, Cu, Fe, Hg, Ni, Pb, Se, Zn), urine (As, Cd, Cr, Co, Hg, Mn, Mo, Ni, Pb, Se), and hair (Hg, Pb).

Selection of appropriate methods of analyses for each situation is also important. Identification and quantification of the causative metal in cases of acute intoxication require rapid and sensitive analytical methods for quick results. For example, significantly increased levels exceeding threshold values for Cd, Cr, Cu, Hg, Ni, Pb and Zn in sewage sludge would require immediate exclusion of such materials from being used as fertilizers. Determination of concentrations (and ratios) at ppm and ppb levels in metal-accumulating species helps in the identification of the typical distribution of what is termed as fingerprint and differences between polluted and non-polluted areas. Biological and environmental monitoring, e.g., determination of Cd, Cr, Ni, Pb, etc. in blood and urine of occupationally-

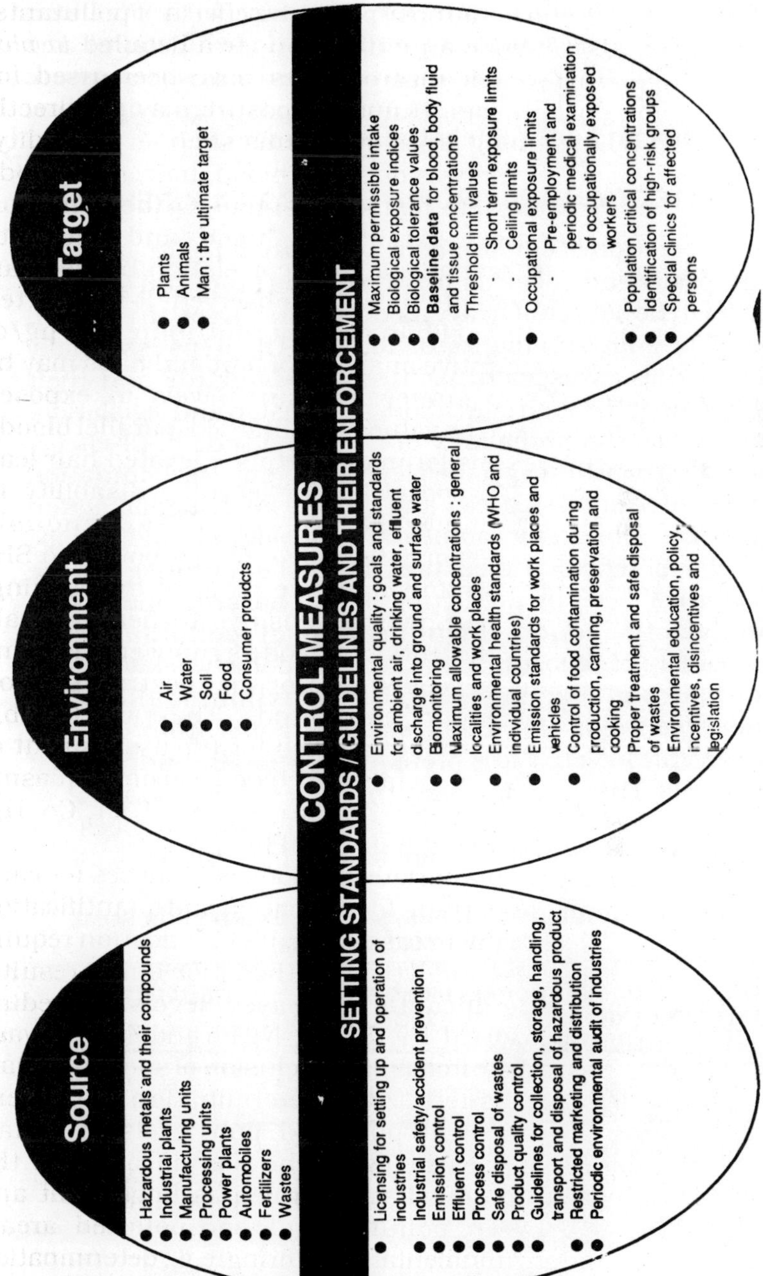

Fig. 3.2. Heavy Metal Pollution : Pathways and Control Measures

exposed individuals calls for accuracy with precision techniques. Assessment of nutritional, behavioural and geochemical influences also warrant the use of very sensitive and sophisticated methods of analyses, stringent dust-free conditions and adequately trained personnel.

3.3 PROTECTION AND PROPHYLACTIC MEASURES

Protection of environment and human health requires appropriate control measures at various points on the path between the source of heavy metal and its target, the human body. These are summarized in Fig. 3.2. The steps essentially involve risk assessment, setting up standards and their enforcement through education, persuasion and legislation. Risk means the probability of the magnitude of ill-effects (including death) at specified levels of exposure. Population studies identify risk areas and groups for remedial action plans. The aim of environmental toxicology should now be to develop remedial measures from the 'curative' to the 'preventive' level by predicting health hazards and taking prophylactic action. Identification of pollutants and assessment of exposure limits (average daily/weekly intake, occupational exposure limits, maximum allowable concentrations, threshold values, population critical concentration, etc.) are prerequisites for initiating such measures. This can be done through a judicious use of bioindicators. Some of the commonly used indices for various metals/metalloids are shown in Table 14.

TABLE 14 GUIDELINE VALUES/BIO-STANDARDS FOR SOME METALS/METALLOIDS

METAL/ METALLOID	OCCUPATIONAL EXPOSURE LIMIT[a] (mg/m^3)	BIOLOGICAL EXPOSURE LIMIT[b]	REMARKS
(1)	(2)	(3)	(4)
Aluminium	2-10	Plasma : 10 μg/l Urine : 200 μg/l	Max. conc. at work place for Al metal, oxide, hydroxide and oxide fume as fine dust: 6 mg/m^3
Antimony	0.5	NA	Classified as carcinogen Max.

(1)	(2)	(3)	(4)
			conc. for stilbine : 0.5 mg/m^3
Arsenic	0.2	Urine : 15-40 μg/1 Hair : <1 μg/g	Classified as carcinogen. Max. conc. for arsine at work place : 0.05 ppm and 0.2 mg/m^3. Technical guideline conc. for As compounds in total dust :0.2 mg/m^3. Sodium arsenite as herbicide is forbidden in many countries. Soil conc. should not exceed 20 mg/kg.
Beryllium	0.002	NA	Classified as carcinogen. US standard for max conc at working area : 2 μg/m^3 of air, 25μg/m^3 for any single determination and 0.01μg/m^3 as one month average. European Community prohibits Br containing cosmetics.
Cadmium	0.05	Blood : 5-15 μg/1 Urine : 1.5 μg/1 Hair : < 1.5 μg/g	Classified as carcinogen. US-Environmental Protection Agency lists Cd as a hazardous air pollutant. In Sweden and Denmark, Cd processing and use (with some exceptions) has been banned. Tolerable weekly

(1)	(2)	(3)	(4)
			intake 0.4-0.5 mg/ person.
Chromium	0.5 (except Cr VI compounds for which OEL is 0.05)	Urine : 15 μg/g creatinine (data from Commission of European Communities)	Not allowed to be added to foodstuffs or medicine in any country. In Germany, max. conc. in soil : 100 mg/kg (loading 4-6 kg per hectare per year for average agricultural use). Classified as carcinogen.
Cobalt	0.5-0.1	Blood:0.65-1.27 μg/l Urine:0.50-1.7 μg/l	Classified as carcinogen. Technical guiding concentration in total dust : 0.5 mg/m^3.
Copper	0.1-1	Liver:> 250μg/g on dry basis	Max. conc. in agricultural soils : 100 ppm or 2-3 kg per hectare per year.
Lead	0.10-0.15	Blood:150-300μg/1 Urine:150μg/g Creatinine. Aminolevulinic acid excretion (ALA-U) : 15μg/1 (the limit for women <45 yr: 6μg/1) Hair : 20μg/1	In Germany paints containing > 1% Pb are prohibited for use inside living areas and warning labels on tins (containing more Pb than this limit) are mandatory. CEC guideline for Pb in gasoline : 0.15-0.40 g/l. In many industrialized countries, pre-employment and periodic medical examination of workers for occupational Pb exposure is obligatory.

(1)	(2)	(3)	(4)
			Tolerable weekly intake : 3 mg/person.
Manganese	1-5	Urine :50 µg/1	Because of lower particle-size, smoke is more dangerous than dust.
Mercury	0.01-0.1	Blood : Inorganic compounds 50 µg/l Organic compounds 100 µg/l Urine : 5–20 µg/l Hair : 50–125 µg/g	Methyl Hg is the most toxic form. OEL for methyl Hg in pregnant females : Nil. Long term daily intake of 3-7 µg/kg as methyl Hg may account for 200-500 ng/ ml in blood or 50-125 µg/g in hair. Tolerable weekly intake : total Hg 0.3 mg and methyl Hg 0.2 mg/ person.
Nickel	0.1-1	Normal range : Blood : 0.05-1.05µg/1 Urine : 0.5-6.1 µg/1	Classified as carcinogen. No MAK limitations. Handling requires extraordinary caution and protection measures. Technical guiding conc. : 0.7 mg/m^3.In the U.S.S.R., max. allowable conc. in waste water directed to biological treatment : 1mg/1
Platinum	0.002-1	NA	German MAK for Pt :0.0002 mg/m^3. Similar US threshold limit values. Standard

(1)	(2)	(3)	(4)
			required not so much for their toxicity as for their sensitizing effects.
Selenium	0.2	Blood : 400 ng/g Urine : 5-30 μg/1	WHO values for safe intake (μg/day) in children : 1-3 yr : 20- 80 4-6 yr : 30-120 > 7 yr : 50-200
Silver	0.01-0.1	NA	Swiss safe-limit for making drinking water bacteria-free: 200 μg/1
Thallium	0.1	Blood : 2-10 μg/1 Urine : 50-300 μg/l	US max.acceptable intake:37 μg/day
Thorium	NA	NA	Exposure limit for ^{232}Th: Ingestion : 3 x10^4 Inhalation : 1 x 10^2
Tin	Inorganic : 2 Organic : 0.01	NA	Momentary (5 min) values > 0.2mg/m^3 may cause local irritation. Daily intake should not exceed 15 ng.
Uranium	0.2	NA	Exposure limit for ^{238}U : Ingestion : 5 x 10^5 Inhalation : 3 x10^4
Vanadium	0.05	Blood : 50 μg/1 Urine : 50 μg/1	Short-term exposure (30 min) should not exceed 0.1 mg/m^3.
Zinc	1-5	Blood : 10 μg/1 Hair:150-200 μg/g	WHO max.tolerable intake : 0.3-1 mg/kg/day. German limit for composted sewage sludge : 300 mg/kg of soil. Zinc chromate is carcinogenic.

(a) Occupational Exposure Limits (OEL) calculated from maximum admissible time-weighted concentrations for an 8-h work-day and a 40-h work week. Data from the American Conference of Government Industrial Hygienists (ACGIH).

(b) Biological Tolerance Valuels (BAT) recommended by MAK or Federal Health Agency, Germany. Upper limits indicate significantly increased concentrations at which adverse effects cannot be ruled out. If increased exposure persists, the source must be identified, reduced or eliminated.

(c) Data from Commission of European Communities (CEC)

Source : Merian, E. (1991) : *Metals and Their Compounds in the Environment*, VCH, Weinheim, Germany.

The protection of human environment is a huge task involving global participation. Many national and international agencies are directly engaged in this activity, namely, the World Health Organization (WHO), United Nations Environment Programme (UNEP), International Register of Potentially Toxic Chemicals (IRPTC), Food and Agricultural Organization (FAO), International Agency for Research on Cancer (IARC), Organization for Economic Cooperation and Development (OECD), US Environmental Protection Agency (EPA), US National Insitute of Occupational Safety and Health (NIOSH), German MAK Commission, Britain's Advisory Committee on Toxic Limits (ACTS), and others. In India, in addition to a separate Ministry for Environment, Forests and Wildlife at New Delhi, we have the National Institute of Occupational Health at Ahmedabad, Industrial Toxicology Research Centre at Lucknow, Schools/University Departments of Environmental Sciences, several voluntary bodies and learned societies (e.g. Academy of Environmental Biology, Indian Association for Occupational Health, Environmental, Mutagen Society of India, Indian Lead Zinc Information Centre, Society for Environmental Communications, Environmental Health Research Organization, etc.), Pollution Control Boards, Ganga-Yamuna Action Plans, education and extension services and legislative measures (e.g., Environment Protection Act, Prevention and Control of Pollution Act,) to deal with such problems. The Government of India has made environmental audit compulsory for all industries through a recent order issued in March

1992. Under the new order, all industrial units would now be required to submit an annual environment audit report to the State Pollution Control Boards stating the amount of pollutants and wastes (both fluid and solid), manner of disposal of such wastes, measures taken and devices installed to curb pollution of air and water, etc. This is certainly a welcome step which will hopefully go a long way in keeping environmental pollution and associated health hazards under control.

CHAPTER 4

HEALTH EFFECTS

4.1 BIOCHEMISTRY AND METABOLISM

4.1.1 Arsenic

The absorption of arsenic depends upon the type of the compound, its solubility and its physical form. Inorganic arsenic is more readily absorbed than organic arsenic and pentavalent form more than the trivalent form. Dermal absorption, however, is more rapid for trivalent than for pentavalent form. Inhalation usually involves particles containing inorganic arsenic. Most of the ingested (95% of trivalent As and 80% of sea-food As) or inhaled arsenic is absorbed through the gastro-intestinal or the respiratory tract.

After absorption, 95-99% of the arsenic binds to globin of haemoglobin in the erythrocytes. Then it is transported to other tissues and body fluids within 24 h. Its deposition occurs in bones, hair, nails and skin where it may persist. Placental transfer of inorganic arsenic has been reported in both humans and experimental animal models. In a study on rats, fetal blood showed levels of arsenic comparable to those of the mother.

Detoxification processes include oxidation of trivalent compounds to pentavalent forms, followed by methylation by methanobacteria forming monomethyl arsonic acid (MMAA) and dimethylarsinic acid (DMAA). In liver, biomethylation occurs mainly by the enzymatic transfer of the methyl group from S-adenosylmethionine. It has been proposed that the reaction between As^{3+} and CH_3^- is very unlikely to occur in aqueous solution. The following possibility of a displacement reaction has been suggested:

$$CH_3^- + As\,(OH_3) \rightarrow CH_3\,As\,(OH)_2 + {}^{\bullet}OH$$

The important examples of interaction of arsenic with other elements include the replacement of phosphorous, negative interaction with iodine and reaction with sulphydryl groups of thiol-containing enzymes. The compounds may also react with thiol groups in the biological systems. Trivalent arsenic compounds may form stable bonds, with thiol groups in the enzymes, disrupting enzyme functions. Alkylarsines and dialkylarsines are formed by reductive methylation of inorganic arsenic and methylarsinic acids which easily condense with SH- groups to form alkylbis (organylthio) arsenic:

$$R\,AsX_2 + 2\,HSR^1 \text{ -> } R\,As\,(SR^1)_2 + 2\,HX$$

Most (70-80%) of the ingested arsenic is eliminated within a week, mainly through the urine as MMAA and DMAA. However, other excretory routes include faeces, skin and hair. Urinary analysis serves as a reliable index of exposure to arsenic.

4.1.2 Beryllium

The absorption of beryllium from the gastro-intestinal tract is very poor (0.01%). Dermal absorption is also limited but inhalation constitutes a serious health hazard as substantial absorption may occur through this route. Experimental studies indicate that about 80% of the ingested metal passes into the faeces unassimilated and about 1% of the dose is excreted in urine. The metal is transported through the blood as collodial phosphate or hydroxide complexes and may enter the cells by mechanisms involving cellular damage by lysozomal enzymes. It resides in the liver until excreted or translocated to bones. Retention is also reported in lungs and skeleton.

4.1.3 Bismuth

The Bulk of the orally administered bismuth remains unabsorbed and excreted in the faeces. Some mucosal uptake is also reported. Plasma concentrations peak within 15-60 min of ingestion. The absorbed bismuth fraction is normally excreted in urine. Considerable renal accumulation has been observed to occur after its parental administration.

4.1.4 Cadmium

Gastro-intestinal absorption of cadmium occurs through passive diffusion and is very poor. Ingested cadmium is absorbed in small intestines (duodenum and ileum), deposited mainly in the liver, spleen and kidneys and excreted in urine (20%), bile and faeces (80%). Cadmium-uptake is related to body Ca-status, i.e., it is strongly inhibited by calcium-uptake. The pattern and extent of the deposition of inhaled cadmium depends on the particulate size (diameter 10-20 μm : nose and pharynx; diameter 5-10 μm tracheo-bronchial region; diameter 5μ m alveolar region). The metal is predominantly bound to erythrocytes in the blood. After absorption from the lungs and the gut, cadmium is transported via blood to liver and other body stores. It crosses the blood-brain barrier and is retained in the brain, though in low quantities. The placenta and mammary glands act as barriers to the transport of cadmium to the foetus and the new born. Its concentration in the new-born is therefore extremely small.

Cadmium has a long biological half-life (>10 years) and its concentration in the body increases with age. The element reacts with many macro-molecules and organic compounds of biological importance, e.g., purine, pyrimidines, nucleosides, nucleotides, RNA, DNA, enzymes, etc. It competes with zinc to inhibit the SH-group of the thiol-containing enzymes. Cd-cysteinate, Cd-cysteinate $(OH)^{--}$, albumin -Cd complexes and aquo-Cd-complexes all exits in the blood plasma. Albumin prevents the production of low molecular-weight fractions. There is convincing evidence to suggest that cadmium regulates the synthesis of mRNA's specific for metallothioneins (MT). Within the tissue, cadmium is mostly found bound to MT. There is a lag period of 3-4 h preceding the synthesis of MT, during which cadmium is distributed between the sub-cellular organelles and high molecular-weight cytosol proteins. Its biosynthesis occurs in response to cadmium as a protective detoxifying mechanism. As a result of such binding, the metal is rendered indiffusible and prevented from binding with proteins and enzymes. Considerable literature exists on the various biochemical effects of cadmium. Table 15 presents a summary of these effects.

TABLE 15 SELECTED BIOCHEMICAL EFFECTS OF CADMIUM

S.No.	Biochemical site	Effects
01.	Hepatic drug metabolizing systems	Contents: ↓ Activity: ↓
02.	RNA synthesis	*In vitro:* ↓ *In vivo:* ↑
03.	Protein synthesis	*In vitro:* translations: by *in vivo* exposure: ↓ by *in vitro* addition: ↓
04.	cAMP metabolism	Levels of cAMP and activity of synthetic enzymes: by *in vivo* exposure: ↑ by *in vitro* addition: ↓
05.	Oxidative phosphorylation	Uncoupling with ↑ state IV respiration *in vitro* and *in vivo*.
06.	Catalase	↓
07.	Carbonic anhydrase	↓
08.	Phosphatases	Generally ↓ to a small extent
09.	Vitamin-D metabolism	↓ by low levels
10.	Phagocytic activity of macrophages	↓
11.	Aldolase	↑
12.	Phosphorylase	↑
13.	Pyruvate carboxylase	↑

14.	PEP carboxykinase	↑
15.	Fructose 6-di-phosphatase	↑
16.	Insulin secretion	↓

Source: Smith, H.A. Cadmium In: *Biochemistry of the Essential Ultra Trace Elements* (ed. Frieden, E), p.361, Plenum Press, New York, 1984.

The complex (MT + Cd) undergoes glomerular filtration but is almost completely reabsorbed in the proximal renal tubules. When tubular reabsorption is impaired, it is excreted through urine. In general the renal excretion of cadmium is minimal and the metal is retained in the kidney.

4.1.5 Chromium

The absorption, retention and excretion of chromium depends on its ionic form. Inorganic Cr^{3+} is poorly absorbed. When linked to amino acids or other biomolecules, the element is more readily taken up by diffusion across the plasma membranes. Such organic complexes constitute the most efficient supply of chromium for man. Absorption of the trivalent form is 1% or less while Cr^{6+} is better absorbed. The latter form readily passes through the erythrocyte membrane and binds to haemoglobin. While the absorption of this metal in the elderly or in the maturity-onset diabetics is not significantly different from that in normal individuals, two- to four-fold higher chromium absorption is reported in juvenile and insulin-requiring diabetes. The primary site in gastro-intestinal absorption appears to be the jejunum. Occupational exposure of workers might occur through inhalation or skin contact with Cr^{6+} in welding plants. Aleveolar macrophages and epithelial lung fluid provide protection against its carcinogenic forms. The reduction of Cr^{6+} takes place in these cell types to produce Cr^{3+} extra-cellularly, thus preventing chromium from entering the lung cells in oxidized toxic forms. Diffusion through skin may also occur which depends on the nature of the compound. For a few compounds the diffusion order is summarized as follows:

Chromates > Chloride > Sulfate > Nitrate

The permeability of membranes to Cr^{6+} allows more chromium to enter and be reduced, resulting in a greater accumulation of the immunologically active chromium in the cell than that absorbed from direct contact with Cr^{3+} compounds.

After absorption, Cr^{3+} binds itself to the β-globulin fractions of serum proteins, especially transferrin. Cr^{6+} accumulates inside erythrocytes where it is reduced to Cr^{3+}. This reducing ability of erythrocytes is an important mechanism in the detoxification of chromium. Transportation of chromium by blood to different organs and tissues occurs in protein-bound forms. Under normal conditions, transferrin is only 30% saturated with iron. The affinity of Cr^{3+} for transferrin is close to that of Fe^{3+} and, therefore, it competes with iron for the same binding sites. Hence, patients with haemochromatosis (with > 50% Fe saturation) retain less chromium than normal or iron-depleted subjects. A relatively higher concentration of chromium occurs in nuclear proteins. Within the tissues, chromium is associated with RNA in a non-dialiasable form. A chromium-containing nucleoprotein has been isolated from beef liver. Unlike most other trace elements, the tissue levels of chromium decline with age. Conversion of inorganic chromium to biologically active forms is essential for its physiological function. Though the exact structure of such forms is not known, the biologically active form of chromium, isolated from brewer's yeast, reportedly contains nicotonic acid and some amino acids such as glycine, glutamic acid and cysteine. The biological activity can be measured in chromium-deficient rats by (i) potentiation of the action of insulin in the glucose oxidation of adipose tissue and (ii) insolated epididymal fat cells.

It has been hypothesized that a diabetic cannot utilize chromium normally. Such patients are probably unable to convert inorganic chromium to a biologically active form. Stress may also affect chromium metabolism adversely and may exacerbate the symptoms of chromium deficiency.

Tissue chromium concentrations are 10-100 times higher than its blood levels. The latter values are not in equilibrium with tissue stores and thus do not give a true picture of its nutritional status. Analysis of chromium in hair has been suggested as a good and convenient indicator for this purpose. High chromium concentrations occur in the liver, kidneys, lungs and spleen. Storage in liver and kidneys is linked to proteins and smaller peptides (e.g. glutathione), in lungs to inhaled chromium, and in spleen to the debris of RBCs. Its penetration to the human brain and placental transport to foetus have been reported in experimental animals. Its excretion is biphasic and occurs mainly through the renal route in urine. The initial rapid phase corresponding to the clearance of chromium from the blood is followed by a slower phase representing clearance from the tissues. Its marginal excretion occurs through bile, faeces, hair and perspiration. Higher excretion of chromium has been reported in diabetics and after oral glucose load in healthy subjects.

4.1.6 Cobalt

Most cobalt compounds have a valency of 2 or 3. The divalent form is more stable as simple salts while the trivalent form is stable in complexes. Normally, cobalt is ingested primarily in the form of various complexes and metalloenzymes. One such complex, synthesized *in situ* exclusively by microorganisms, is vitamin B_{12} also known as cyanocobalamine. It contains 4.35% cobalt.

The principal routes of its entry into the body are inhalation, oral, skin and unprotected mucous membranes. After oral ingestion, a high absorption (60-90%) of cobalt from the gastro-intestinal tract has been recorded. Its occupational exposure occurs mainly through skin and inhalation. The uptake of cobalt is apparently coupled with that of iron. Anaemic patients responsive to the iron therapy exhibit increased cobalt absorption. As cyanocobalamine, the metal contributes to the formation of red blood cells and is essential to the normal functioning of all cells, particularly those of the bone marrow, nervous and gastro-intestinal systems. Cobalt inhibits cellular respiration and enzymes of the citric acid cycle. Such a hypoxic stimulation, mediated by cobalt, leads to an increased erythropoietin biosynthesis, which in turn, triggers the erythropoiesis in the bone marrow. Storage sites for this metal are bones, liver, spleen, kidneys (the longest time of retention), endocrine glands and lymph nodes. Although inorganic (non-vitamin B_{12}) cobalt is invariably present in organs and body fluids, its physiological role is not known.

Cobalt shows absorption-antagonism with iron and manganese. It may replace Zn, Mg and Mn in a variety of enzymes *in vitro* due to similarities in ionic size. Addition of cobalt *in vitro* has been reported to increase the activity of a variety of enzymes such as alkaline β-glycerophosphatase, dehydrogenase, hydrogenase, nitrate reductase, oxidase, pancreatic amylase and lipase, phosphatase, proteinase and zymosan. It activates the thyroid gland in patients with iodine deficiency and best results are obtained by the combined use of cobalt and iodine.

Excretion of ingested cobalt occurs mainly through faeces (60-80%) and urine (20-30%). It is fast and intensive during the first 24 h and most of it is completed within a week. Beyond that, the metal is slowly excreted in micro-quantities in sweat, saliva, hair, nails and seminal fluid. Inhaled cobalt particles are rapidly cleared from lungs. Tracer studies indicate biphasic renal excretion of inhaled cobalt; the bulk of the excretion occurs with a half-time of 10 days and the remainder with a half time of 90 days.

4.1.7 Copper

Copper is readily available as Cu (I) or Cu (II) in inorganic salts and organic complexes. About 30-50% of the oral intake is absorbed mainly from the duodenal mucosa through a copper-binding protein. The extent of absorption depends on the form of copper, the acidity of gastro-intestinal tract and the presence of other substances in diet. Phytate, high ascorbic acid, certain amino acids and ligands which form complexes with copper, and metals, such as cadmium, mercury, molybdenum, silver and zinc reduce its absorption and alter its tissue distribution.

Absorption may occur through skin. Upto 50 mg of the metal from copper bracelets (traditionally used to alleviate muscular and cardiovascular ailments) may be lost per month. Tracer studies have revealed perfusion of copper complexes through intact cat-skin and the formation of copper glycinate complex under aerobic conditions:

$$Cu^{2+} + 2H_2NCH_2COOH + ½O_2 \rightarrow [Cu(H_2NCH_2COO)_2] + H_2O$$

This suggests probable oxidation of copper by amino acids present in sweat.

Absorbed copper is loosely associated with albumin and is distributed to almost all tissues and body fluids through plasma. Organic complexes permit transport through the cell wall. The metal forms an integral part of the DNA or RNA molecule. Ceruloplasmin in liver, and erythrocuperin in bone marrow are major copper-containing proteins. Its larger amounts are present in brain, liver, heart, kidneys and the pigmented parts of the eye. The key organ of its metabolism is liver where it is stored in soluble fraction (52%), nuclei (15%), mitochondria (12.4%), and microsomes (14%). The metal is a constituent of many enzymes such as hemocyanin, tyrosinase, lysal oxidase, ascorbic acid oxidase, cytochrome oxidase, uricase, monoamine oxidase, dopamine hydroxylase etc. Copper is needed for the production of haemoglobin, oxygen transport and formation of aortic elastin and bone. Transfer of iron from tissues to the plasma requires the oxidation of the ferrous form. This is catalysed by ceruloplasmin. The metal facilitates incorporation of iron into haemoglobin. The primary biochemical effect in bone disorders, neonatal ataxia and cardiovascular dysfunction is a substantial reduction of copper in the cytochrome and amino oxidases.

Unabsorbed copper, as well as a high proportion of injected copper, is excreted through bile and faeces. Urinary excretion is less than 50 μg per 24 h. Negligible amounts may be excreted through sweat, hair and menstrual flow.

4.1.8 Gold

The main source of gold exposure to humans are gold containing drugs. Approximately 25% phosphine containing

Au (I) complexes, e.g., auranofin, which are used as anti-arthritic drugs, are absorbed from the gastro-intestinal tract, reaching peak serum levels within hours. The oligomaric Au thiolates, e.g., myocrisin or solagnol, are given parenterally. In the blood, the thiolate forms of gold react with albumin and other serum proteins involving ligand exchange reactions. The bulk of intravasal gold is protein bound of which more than 85% is bound to albumin. Transport across biological membrane has been observed and the metal may enter the ertythrocytes and other cells.

Gold concentrates in the liver, kidney and the synovial fluid. The metal may persist in the body for prolonged periods. The principal depository is lysosomes which, upon accumulating the metal, become electron-dense and are called aurosomes. Retention in autopsy samples is discernible even after 20 years of cessation of gold therapy. Detoxification mechanisms include reduction of Au (III) to Au (I), possibly by protein disulfides, thiols and thioethers. Interaction of myocrisin with CN have been reported. These are of interest because the cyanide inhaled by tobacco smokers markedly alters gold distribution and metabolism in the body. Interactions may also occur with other antiarthritic drugs such as d-pencillamine.

Gold is mainly excreted through urinary and faecal routes. For orally administered preparations such as auranofin, less than 20% of the clearance is renal, in contrast to over 60% for parenteral preparations of myrocrisin or solagnol.

4.1.9. Lead

Absorption of ingested lead from the gastro-intestinal tract is very poor (5-15%) ; about 293 mg of the average daily intake of 300 mg is excreted. The rest (mean : 7 mg, range : 3-11 mg/day) is mainly retained in the bones. Absorption is affected by a number of factors : (i) *chemical form and solubility* : acetate, chloride, oxide and tetraethyl compounds are readily absorbed; sulfides, sulfates, chromates and carbonates are less soluble but fairly absorbable, (ii) *Diet* : the rate is significantly enhanced under fasting conditions, low calcium and vitamin D, iron deficiency and presence of ascorbic acid, amino acids with SH-groups and bile acids. Elements competing for the same carrier systems in the intestinal epithelium, such as calcium, iron, magnesium, phosphate, ethanol and high-fat diets reduce the absorption of lead, and (iii) *Age* : young children absorb more lead than do adult humans. Pulmonary absorption of inhaled lead is a two-tier process : (i) deposition in the deeper part of the respiratory tract, and (ii) absorption of the deposited fraction. The rate depends on the state of aggregation (gas or solid), particulate size, concentration in the ambient air, respiratory volume and distribution of lead within the respiratory tract. Larger particles are cleared from the upper respiratory tract by ciliated epithelial cells. Substantial absorption occurs from the lower respiratory tract. A portion of the lead retained in alveoli may, however, be cleared by macrophages. Animal studies also provide evidence for some percutaneous absorption but it is not significant.

Tracer studies indicate distribution of lead in three compartments : (i) blood with very short half-life (hours/days), (ii) soft tissues with medium half-life (weeks), and (iii) skeleton representing a pool of very long half-life (months/years). More than 90% of the blood-lead is bound to haemoglobin in erythrocytes. A specific lead-binding protein (molecular weight: 10,000) has been identified and attributed with protective function. The metal in blood is largely exchangeable with tissues/organs; the distribution depends on the time and extent of exposure as well as on the turnover rate for individual compounds. Retention chiefly occurs in bones and teeth (>90%), with moderate content in liver and kidneys where it has been detected in sub-cellular fractions (nuclei, membrane, mitochondira, etc.) also. Though the principal binding sites for lead are macromolecules, mainly proteins with several free sulfhydryl groups (e.g. a metallothionein containing cysteine), it may form less stable compounds with other side chains of amino acids, e.g., E-amino group of lysine, carboxyl group of glutamic and aspartic acids and phenoxyl group of tyrosine. Interactions with cadmium, calcium, copper, iron, magnesium, phosphorus, zinc, vitamin D and several enzymes, particularly in the heme pathway, have been reported. Blood lead levels higher than 30 μg/dl lead to increased concentration of δ-amino-levulinic acid (ALA) due to an inhibitory action of lead on ALA dehydrogenase. Estimation of the activity of erythrocyte ALA has been suggested as an index of lead exposure.

The metal crosses the blood-brain and placental barriers. Within the brain, the highest lead concentrations are found in the cortical grey matter and basal ganglia. Cord blood-lead levels correlate with maternal blood levels. Foetal uptake of lead is apparent by the 22nd week of gestation and increases throughout development. Excretion of unabsorbed lead occurs through faeces, which is the main route of its elimination from the body. Sizable excretion may occur through the sweat. Glomerular filtration, tubular secretion and renal excretion also occur, though to a lesser extent.

4.1.10 Manganese

The absorption of manganese occurs through respiratory and gastro-intestinal tracts. Organo-manganese compounds may also be absorbed through the skin. A part of the inhaled manganese is cleared by mucociliatory action. Absorption of ingested manganese is poor (3-4%). It does not appear to be under homeostatic control and is largely independent of manganese content in the diet or the body burden of this element. The mechanisms of its absorption are not precisely known but it occurs throughout the length of small intestine. Its absorption rate is considerably increased with alcoholism, in iron deficiency anaemia (upto 7%), and during pregnancy and infancy.

The half-life of manganese absorbed orally is about 10 days and that for parenterally injected about 40 days. The absorbed manganese enters the portal blood where most of it gets bound to α_2-macroglobulin. It is also incorporated in erythrocytes during erythropoiesis. In the liver, a fraction of the element is oxidized to Mn^{3+} by ferroxidase I and then bound to transferrin. The organs where manganese is mostly concentrated are liver, pancreas, kidney and the intestines. Within the cells, manganese is found in mitochondria. The metal acts as a cofactor in several metaloenzymes, such as arginase, pyruvate carboxylase, superoxide dismutase, and elicits important functions in the brain, in the formation of connective tissues and bones, and in the metabolism of biogenic amines, carbohydrates and lipids. Manganese catalyses the formation of glucosamine-serine linkages during the synthesis of mucopoly-saccharides of cartilage. The element can penetrate both blood-brain and placental barriers. The half-time for brain is longer than that for the rest of the body.

Variable excretion pattern is reported, which perhaps maintains homeostasis. The excretion of inorganic manganese occurs mainly through the bile and faeces. Its marginal excretion occurs through pancreatic juice and kidney. Only 0.1-1.3% of the daily intake is excreted through urine. However, renal excretion for organo-manganese tricarbonyl compounds, which are used as additives in gasoline, is substantial.

4.1.11 Mercury

The main route of entrance of mercury into the body is through inhalation, as would be expected from its monoatomic nature, high vapour pressure, lipid solubility and high diffusibility. Approximately 80% of the inhaled mercury is retained. The absorption of the inorganic mercury from food is about 7%. In contrast, the absorption of methyl mercury from diet is about 95% in adults. Most forms of mercury can penetrate the skin to some extent; systemic absorption of alkyl mercurials is substantial and poisoning may result from dermal application of such compounds.

The metal is distributed to almost every tissue or organ of the mammalian body, including brain, placenta and foetus. However, its highest concentrations are found in the kidney of an exposed animal. Mercury vapours are rapidly oxidized in tissues to divalent ionic form. Organo-mercurials (e.g. phenylmercury and mercurial diuretics) also undergo rapid transformation into inorganic mercury. Short chain alkyl-mercury compounds (e.g. methylmercury) undergo slow biotransformation; for example, in humans, less than 1% per day of methylmercury is converted into its inorganic form. The toxic properties of methylmercury may partly be attributed to the stability of the C-Hg bond. The ratio of red cells to plasma content depends on the form of mercury; it is highest for methyl mercury and lowest for inorganic mercury. Hair content is a useful index for assessing blood levels of methyl mercury ; the concentration in hair being about 250 times higher than the concentration in blood.

Mercuric cations have a high affinity for sulfhydryl groups. Since many proteins contain SH-groups or disulfide bridges, the metal can cause widespread alteration in their functions. The distribution in the brain and the central nervous system is largely governed by the available mercury binding sites/SH-groups.

The excretion of mercury follows several routes. It can be volatilized from the lungs and skin, can be eliminated via urine, sweat, milk, saliva, intestinal mucosa, bile, faeces, or stored in and shed through hair and nails. The average half-life of methyl mercury in man is about 70 days (range 46-120 days). The principal route of excretion is through faeces, but in cases of prolonged exposure, urinary excretion may marginally exceed faecal elimination. Concentration of total mercury in urine has no correlation with blood mercury levels in populations heavily exposed to methyl mercury.

4.1.12 Nickel

Less than 10% of the ingested nickel is normally absorbed. A higher percentage may be absorbed during pregnancy. Food and beverages reduce or prevent the absorption of Ni^{2+} from the gut. Transfer of Ni^{2+} across the mucosal epithelium involves an energy driven process (rather than simple diffusion). It is connected with the iron transport system located in the proximal part of the small intestine. In humans, approximately 35% of the inhaled nickel is absorbed; the remaining part is cleared out by the mucociliatory action in respiratory passages and either swallowed or expectorated. Its compound, Ni-carbonyl, which is highly volatile, is absorbed readily from lungs. Percutaneous absorption of nickel may also occur.

Transport of Ni^{2+} in plasma is mediated by binding to albumin and unfiltrable ligands. The latter include amino acids (e.g. histidine and cystine) and small polypeptides. An α-macroglubulin (nickeloplasmin), despite being the major Ni- binding protein, does not play a significant role in extracellular transport as the metal so bound is not readily exchangeable with exogenous Ni^{2+}. Nickel carbonyl is lipid soluble and penetrates the blood-brain barrier, but nickel concentration in brain has been reported to be substantially less than that in other tissues. Within the body, it is widely distributed in most tissues and organs. The metal has been consistently reported to be present in RNA. It is a potent activator of several enzymes and probably plays a role as a bioligand in iron absorption, regulation of prolactin, and in the structure and function of membranes. Interactions with calcium, chromium, copper, cobalt, iodine, iron, magnesium, manganese, molybdenum, phosphorus, potassium, sodium and zinc have been reported. Placental transport and appreciable retention in foetal tissue may occur.

Most of the ingested nickel remains unabsorbed and is excreted in faeces. Urinary excretion is small and in the form of low molecular complexes which include the complexes with histidine and aspartic acid. Healthy adults contain significant amounts of nickel in sweat, suggesting its active secretion by sweat glands. Excretion also occurs through milk, hair and exhaled air.

4.1.13 Selenium

The exposure to environmental selenium occurs mainly through inhalation (0.02 mg/g) and ingestion (0.1 μg/g). Its absorption takes place in the duodenal part of the gastro-intestinal tract. Absorbed selenium associates with plasma proteins (e.g. seleno-cysteine and selenomethionine) which carry and deliver it to all cells and tissues. Its higher concentrations are observed in renal cortex, brain, endocrine and reproductive organs. Its incorporation into proteins of liver and erythrocytes has also been observed. About thirteen selenopeptides and selenoproteins have been detected. It is a component of various enzymes such as glutathione peroxidase (GPX), glycine reductase, muscle cytochromes, etc. Studies in rodents indicate that about 40% of the total body selenium is bound to GPX. GPX catalyzes the disintegration of lipid peroxides and, therefore, has important antioxidant function alongwith superoxide dismutase, catalase and tocopherol.

Selenium compounds are metabolized in the body via two pathways : (i) reduction followed by methylation, or (ii) direct incorporation into or binding with proteins. It can cross the placental barrier and may be detected in foetal tissues.

The biological half-life of selenium is reported to be 14.4 days. Excretion occurs through faeces, urine and exhaled air. At higher intake levels, exhalation becomes an important route of excretion. A higher phosphorus, protein and metheonine content in diet also causes an increase in the pulmonary and renal excretion of selenium.

4.1.14 Silver

The absorption of silver is poor in both the gastro-intestinal tract and the lungs, though it can also be absorbed through the skin. A part of the absorbed metal is fixed in tissues. Retention is greatest in the reticulo-endothelial system. Human beings may accumulate a body burden of approximately 9 mg (calculated from average daily-intake of

27 to 88 μg) over a period of 50 years. The half-life of silver in liver and skin is longer than that in other tissues. The silver ion (Ag^+) has a strong affinity for sulfhydryl, amino, and phosphate groups. It forms complexes with amino acids, purines, pyrimidines, nucleosides, nucleotides, phosphates, proteins, DNA and RNA, and exhibits antagonism to copper. Its excretion occurs primarily through faeces (90%) while only traces are found in urine.

4.1.15 Thallium

The absorption of thallium from mucous membranes after ingestion or inhalation is rapid and substantial (upto 80-100%). It may also be absorbed on contact with skin. Though a rapid distribution occurs from blood to tissues, thallium is retained longer in blood plasma (as compared to other metals, such as barium, nickel or zinc), suggesting its cumulative nature. Its storage occurs mainly in intracellular fluid where it combines with amino acids and builds up in bones as phosphate. Tracer studies using ^{204}Tl indicated tropism of the isotope in the following order :

kidneys, testes > liver > spleen > prostate > brain > skin, hair.

Competitive replacement of Tl^+ by K^+ has been reported. Binding of thallium by ion-exchange in the gastro-intestinal tract constitutes an effective mechanism for its detoxification. The bulk of excretion (70-90%) occurs via urine and faeces within four weeks of ingestion. It may also be eliminated via saliva and skin. An analysis of thallium in scalp hair may provide forensic diagnosis and therapy control.

4.1.16 Tin

The absorption of ingested inorganic tin is very poor (<5%). It is influenced by the oxidation state; Sn (II) compounds are more readily absorbed than Sn (IV) compounds. The presence of organic acids in diet enhances the absorption of inorganic tin. There is considerable variation in the rate and extent of absorption between different compounds. As a rule, tin compounds with short alkyl chains are more readily absorbed from the gastro-intestinal tract. Trialkyl tin compounds are well absorbed through the skin. The extent of absorption of tin compounds through inhalation has not been adequately assessed.

The absorbed tin leaves the vascular system rapidly. Despite its low retention in humans, it is detected in most organs and tissues; the accumulation is more in bones than in soft tissues. Other principal sites of deposition are the lungs, liver and kidneys. Penetration through the blood-brain and placental barriers does not appear to be significant. Biomethylation does not appear to be an important pathway for the transformation of organotin compounds and evidence for microbial methylation of inorganic tin is so far equivocal. Dealkylation and dearylation of tetra-, tri and di-substituted organotin compounds occurs in the liver. The dealkylation of diethyl tin compounds takes place in the gut and other tissues. Interferences with porphyrin and haeme biosynthesis, mitochondrial oxidative phosphorylation, activity of δ-aminolevulinate dehydratase and riboflavin, and metabolism of calcium, copper, iron, selenium and zinc have been reported.

The clearance of most organotin compounds is slow and biological half-life appears to be longer in brain than in other organs. Unabsorbed Sn (>90%) is excreted in faeces, and absorbed part is excreted mainly in urine. It may also appear in bile, pancreatic juice, milk and sweat. The fraction excreted with bile varies with the type of compound. Ethyltin trichloride is mainly eliminated with urine; diethyltin is excreted through all three routes, viz., bile, faeces and urine; while triethyltin may leave the body via urine and milk.

4.1.17 Vanadium

The absorption of vanadium inhaled in particulate forms entering the lungs is rapid compared to its oral absorption. Based on the tracer studies data, the absorption period of the metal is given as follows. Oral and subcutaneous: 1-6 h, intraperitoneal : 30 min, and intratracheal : 5 min. The absorbed vanadium is transported in serum mainly bound to trasferrin. Most of the vanadium entering the blood stream is rapidly eliminated with only a small fraction retained and deposited in tissues. Isotopic studies in rats reveal distribution of vanadium in the following order: bones > kidneys > liver > spleen > intestines > stomach > muscles > testes > lungs > brain. It occurs intracellularly in the vanadyl (IV) form and is associated with high molecular weight proteins in the soluble fraction of hepatic cells. Subcellular distribution in liver is as follows : nuclei and debris (43%), mitochondria (37%), microsomes (10%) and with the supernatant fluid (11%). Much of the vanadium present in physiological systems is bound to coordinating groups, such as -SH, -SS, -OH, -N-, -COO-, and -PO_4. It affects several enzymatic reactions, the sodium pump, oxidation-reduction processes, metabolism of iron, carbohydrates and lipids, bone and teeth formation and hormonal functions. The bulk of ingested vanadium remains unabsorbed and is excreted in faeces, while the remainder is eliminated via urine and bile.

4.1.18. Zinc

The absorption of ingested zinc takes place in small intestines, particularly through the distal duodenum and proximal jejunum. It is equally well absorbed as oxide, carbonate and free metal, but poorly as sulphate or as mixed oxides of zinc, iron and manganese. The rate of absorption shows considerable variation (20-80% from food, 43-66% from drinking water) as it is influenced by several factors : (i) zinc status: absorption is more in deficiency states and less when zinc status is adequate or high, (ii) pathological conditions depress absorption, (iii) diet : interfering substances in food, e.g., high phytate, fibre, soya protein, cadmium, calcium, copper, etc. reduce zinc absorption, (iv) enhancers of zinc absorption are only a few, e.g., red wine (but no other alcoholic beverage), histidine and other amino acids, and (v) increased availability from mother's milk as compared to cow's milk has been suggested but not proved beyond doubt. Metallothionein plays an important role in the homeostatic regulation of zinc absorption and metabolism. The absorption probably follows a saturable carrier-mediated process.

Absorbed or parenterally administered zinc binds to plasma proteins (albumin, α-2 macro-globulin and transferrin) and is distributed throughout the body tissues. Its highest concentrations are reported in prostate, liver, muscle and kidneys. There is a rapid uptake and turnover rate of retained zinc for some vital organs, such as pancreas, liver, kidney and spleen. Bones and the central nervous system exhibit a slower uptake; the metal remains bound for a longer time in these tissues. The metal crosses the placental barrier to reach the foetus. Within the body, the metal is a constituent of a large number of enzymes and proteins and plays an important role in enzymatic reactions, nucleic acid metabolism, protein synthesis, maintenance of membrane structure and function, protection against free-radical damage, replication, transcription and translation of genetic material, and hormonal activity.

Tracer studies using ^{65}Zn revealed a two-phased whole-body elimination kinetics: 10-20% with a median half-life of 7-14 days and 80% with half-life ranging between 160-500 days. Unabsorbed zinc and a small fraction of endogenous zinc is excreted through pancreatic juice, bile and faeces. Faecal elimination (5-10 mg/day) accounts for 70-80% of ingested zinc. Normal urinaty excretion of zinc is low (0.4-0.6 mg/day). It is bound mainly to amino acids and porphyrins. Elimination through this route increases under certain conditions and also following administration of EDTA. Significant quantities of zinc may be lost in sweat (upto 3 mg/day), especially in hot weather.

4.2 EFFECTS ON ENZYMES AND SUB-CELLULAR ORGANELLES

Metals elicit essential regulatory and toxic effects by binding with receptors and protein molecules at cellular and subcellular sites and through innumerable enzymatic reactions in the body. They may form structural components of metalloenzymes or serve as cofactors, activators or inhibitors of enzymes and proteins (Table 16). Metallothionein (MT) plays an important role in the storage and transfer of certain metal ions. Metals ions, which may induce MT synthesis in cultured cells or *in vivo*, include bismuth, cadmium, copper, gold, mercury, silver and zinc. This ubiquitous protein is a fast responding homeostatic mediator which regulates the flow of metal ions through the cells. It participates as a metal donor or acceptor in the biosynthesis and degradation of metalloenzymes and metalloproteins. The relative binding affinity of MT for some heavy metals is as follows :

$$Hg > Cu > Cd > Zn$$

Metals may act as systemic toxicants by affecting vital enzymatic and other regulatory processes. Similarly, ligand binding properties of essential nutrients or competitive interactions between different metals may alter their uptake and metabolism which may ultimately have serious implications. For example, cadmium, cobalt, nickel, tin and zinc replace the magnesium ion in a number of enzymes involved in phosphate transfer. This is significant because the toxicity of metals in organisms has been reported to be inversely proportional to the concentration of magnesium in the medium. Many metals (e.g., arsenic, mercury, selenium and thallium) inhibit the SH-group of aminoacids, peptides, proteins and enzymes (e.g., glutathione, coenzyme A, cysteine, homocysteine, ergothioneine, some cytoplasmic and membranous proteins) with disruption of important biological functions. Toxic metal compounds may alter redox processes or even directly generate reactive oxygen species by various mechanisms (please see section 4.3).

Complex formation of metals with DNA constitutes the most serious reaction from genetic point of view. Induction of cross-links between both the strands of DNA has been reported following exposure to cobalt, copper, manganese, platinum and zinc. Other mechanisms include chelation (silver, mercury) and binding with the phosphate group stabilizing the DNA helix (cadmium, cobalt, copper, manganese, nickel, zinc). Metals which may induce chromosomal damage in human cell cultures or *in vivo*, include arsenic, cadmium, chromium, lead, mercury, nickel and zinc. The understanding of underlying reactions is important in the context of their possible mutagenic, carcinogenic, or teratogenic effects. Some metal ions are concentrated in nucleus, nucleolus, mitochondria and nuclear proteins. High levels of metals in mitochondria are particularly interesting as in these cell organelles reside the multi-enzymatic system related to respiratory chain, citric acid cycle, nutrient degradation and energy releasing processes. It is suggestive of the crucial role of these metals at cellular and sub-cellular levels, both from the physiological and toxicological view point. Thus, Cr (VI) and Cr (III) may cause intra-nuclear free-radical production or formation of condensation polynucleotides by binding to guanine and the sugar part of DNA. Iron, which forms the "energy backbone" of vertebrate physiology, is also involved in DNA synthesis and is responsible for catalytic effects of cytochromes, catalases, peroxidases and other mitochondrial enzymes; lead interferes with several enzymes that participate in haeme synthesis, e.g., stimulation of the mitochondrial enzyme, δ-aminolevulinic acid synthetase, inhibition of cytosolic enzymes, δ-aminolevulinic acid dehydratase and ferrochelatase; manganese acts as a cofactor in the activation of many mitochondrial enzymes; the neurotoxic effects of methyl mercury have been linked to its high concentrations in nuclear, microsomal, mitochondrial and myelin fractions of brain cells; specific translocation of cations across the synaptosomal or mitochondrial membranes and interference with Ca^{2+} or Na^{2+} / K^{+} ATPase systems; selenium interacts with B and T lymphocytes, macrophages and natural killer cells of the immune system; tellurium increases the frequency of DNA breaks; organotin compounds bind to mitochondria and interfere with oxidative phosphorylation; vanadium inhibits coenzymes A and Q, with important roles in lipid metabolism, mitochondrial respiration and bioenergetic mechanisms; zinc plays a vital role in varied enzymatic processes, DNA and RNA metabolism and protein synthesis; and zinc enzymes are involved in replication, transcription and translation of genetic material in all species. The above examples are obviously illustrative and not exhaustive.

TABLE 16 EFFECTS OF SOME METALS ON CELLULAR TARGETS

Metal	Intra-Cellular Distribution/Major Targets of Action	Metalloenzyme/ Metalloprotein	Effect on Enzymes/Proteins	
			Elevation/ Activation	Depletion/ Inhibition
(1)	(2)	(3)	(4)	(5)
Arsenic	Damage to DNA without causing gene mutation.	Arsenobetaine, Arsenocholine-O-phosphatidyl-trimethyl arseni-umlactic acid	Arginase	Alkaline phos-phatase, Adenosine-tri-phosphate (ATP). Replaces phosphorus in the phosphate group of desoxyribonuclic acid (DNA), Sulfhydryl (SH)-group of essential enzymes.
Beryllium	Lysosomes, nucleus/ nucleolus. Cell disrup-tion of macrophages with release of Be ion into surrounding tissues.	Colloidal beryllium phosphate and hy-droxide complexes with plasma globulins.	α-and β-globu-lins, Immuno-globins A and G (1g A and 1g G)	A large number of enzymes including those involved in DNA synthesis and catalysis of of energy processes. Albumin, Collagen, Enolase,

(1)	(2)	(3)	(4)	(5)
				Glycosaminoglycans, IgM, Lactate dehydrogenase, Phosphorylases, Phosphoglucomutase, Phosphoglyceric aldehyde, SH-group proteins, Transferrin, etc.
Cadmium	Finger-loop domain of gene-regulating proteins. Elongation phase of DNA synthesis.	Albumin-Cd-complex, Aquo-Cd-complex, Metallothionein	Aldolase, Ceruloplasmin, Cyclic adenosine-monophosphate (cAMP), Fructose -6-diphosphatase, PEPcarboxykinase, Phosphorylase, Pyruvate carboxylase	Adenylate cyclase, Alkaline phosphatase, Carbonic phosphatase, Carboxypeptidase (peptidase activity abolished, no effect on esterase activity) Catalase, Choline esterase, Glutathione transferase, Hydrocarbon hydroxylase, O-methylguanine-DNA-methly-transferase Phosphatases, SH-group of essential

(1)	(2)	(3)	(4)	(5)
				enymes, Zinc enzymes (competition for binding sites)
Chromium	Associated with nucleoproteins and RNA in a non-dialysable form. Breaks DNA by direct radical formation and/or through inhibition of DNA polymerases, glycine, cysteine and glutamic acid. Complex with transferrin fraction of β-globulins.	Glucose tolerance factor (GTF) : an organic complex containing chromium, nictotinic acid, glutathione or its constituent amino acids.	Insulin	DNA polymerase, Transferrin, Triglycerides
Cobalt	Addition to human leucocytes and diploid fibroblasts causes a reduction in mitotic index with no effect on mutation.	Albumin-Co-complex, Cyanocobalamine, Diol dehydrase, Glutamic isomerase, Transmethylases.	Alkaline β-glycero-phosphatase, Dehydrogenases, Heme oxygenase Lipase, Nitrate reductase Oxidases, Pancreatic amylase, Phosphatases, Proteinases, Zymosan.	Ctyochrome P-450, Haeme, Transferrin.

(1)	(2)	(3)	(4)	(5)
Copper	Distribution in hepatic cells: Soluble fraction 52% Nuceli 15% Mitochondria 12.4% Mircrosomes 14%	Ascorbic acid oxidase, Ceruloplasmin, Cysteine-Cu-complex, Cytochrome oxidase, DNA, Dopamine hydroxylase, Erythrocuperin, Galactose oxidase, Hemocyanin, Histidine-Cu-complex, Lactocuperin, Lysyl oxidase, Metallothionein, Monoamine oxidase, Phenol oxidase, RNA, Superoxide dismutase Tryptophan deoxygenase Tyrosinase, Uricase.	Glucose -6-phosphate dehydrogenase, Lysyl oxidase	Glutathione -S-transferase.
Gold	Aurosomes (lysosomes with deposits of gold) Cytosol.	Albumin-Au-complex, Globulin-Au-complex, Metallothionein.	Immunoglobulins	Acid phosphatase, Aspartate transcarbomylase β-N-acetylgluco-saminidase Elastase, Glucuronidase Hyaluronidase,

(1)	(2)	(3)	(4)	(5)
				Lysozymes, Neutral proteases, Nucleoside mono-and di-phosphokinase, Prostaglandin, Pyruvate oxidase, Tyrosinase.
Lead	Haemoglobin Causes chromosomal aberrations and breaks in DNA strands	--	δ-aminolevulinic acid synthetase, Hexokinase	Enzymes responsible for calcium-dependent reactions, porphyrin and haeme-synthesis. Calmodulin, Carboxypeptidase [peptidase activity abolished, no effect on esterase activity], Choline esterase, Ferrochelatase, Glutathione peroxidase, Immunoglobulins, o^6- methylguanine-DNA-methyltransferase, Phosphobilinogen synthetase.

(1)	(2)	(3)	(4)	(5)
Manganese	Concentrated in mitochondria and nuclei, Nucleotides, DNA, RNA.	Arginase, Complexes with catechols, ethanolamine and porphyrin, Lectins, Malic enzymes, NADH Nitroreductase, Pyruvate carboxylase, Superoxide dismutase.	Alkaline phosphatase Arginase, ATPase, Biogenic amines, cAMP, CoA carboxylase, Farensyl pyro-phosphatase, Glycyerol transferase, Insulin, Peptidases, Phosphophenyl Pyruvatecar-boxykinase, Polymerases, Prolidase, Pyruvate carboxylase, Succinyl dehydrogenase, Superoxide dismutase.	Adenylate cyclase, Dopamine Magnesium enzymes/ proteins.
Mercury	High concentration in Purkinje cells of cerebellum and certain neurons of spinal cord, medulla and mid-brain.	Metallothionein	Hydration and hydrolysis of vinyl-ether-linkage of plasmogens (1-alkenyl,	Carboxypeptidase (peptidase activity abolished, no effect on esterase activity), Glutathione peroxidase

(1)	(2)	(3)	(4)	(5)
	Distribution of methyl mercury in different fractions: Supernatant nuclei > microsomes > synaptosomes > mitochondria > myelin.		2-acyl-glycyrol phospholipids) Orinthine de-carboxylase in fotal brains.	Glutathione-s transferase, 0^6-methylguanine -DNA-methyl transferase, SH-group of essential enzymes, Disulfide bridges in proteins, tRNA cycles.
Nickel	Role in structural organization of RNA.	Complexes with albumin, and histidine-rich proteins, Nickeloplasmin, Urease.	Acetyle CoA-synthetase, Alanine amino-transferase, Arginase,Aspartate, amino transferase, Carboxylase, Dehydrogenases, Glutathione, Transaminase, Trypsin.	Acid phosphatase, Benzopyrene hydroxylase, Glutathione peroxidase, Glutathione reductase, Hepatic glutathione.
Selenium	Cytosol	Dehydrogenases, Factor-3 (hepato-protective), Glutathione pero-oxidase, Glycerl reductase, Muscle	Coenzyme Q/ Ubiquinone, Cytochrome P-450, Formate dehydrogenase Glycine reductase,	Hepatic urease, Metheonine adenosyl transferase, SH-group enzymes, Succinic dehydrogenase.

(1)	(2)	(3)	(4)	(5)
		cytochromes, Selenocysteine, Selenotaurine.	Hemeoxygenase Nicotinic acid hydrolase, Xanthine dehydrogenase.	
Silver	Bound to G-C rich regions of DNA	--	--	Ceruloplasmin, Glutathione peroxidase.
Thallium	--	--	--	Albumin: globulin coefficient, Alkaline phosphatase, SH-group enzymes.
Tin	--	Gastrin	Heme oxygenase, 1gG, 1gM.	Acid and alkaline phosphatases, Aromatic hydroxylase, Zinc metalloproteins.
Vanadium	Sub-cellular distribution in liver: Nuclei : 43% Mitochondria : 37% Microsomes : 10% Supernatant : 10%	Transferrin	Insulin (?)	Cytochrome P 450, Monoxygenase, Triglycerides.

(1)	(2)	(3)	(4)	(5)
Zinc	Concentrated in nucleus and mitochondria; Involved in biosynthesis of proteins.	Constituent of more than 200 proteins/enzymes in all species. Required for about 100 mammalian enzymes. Zinc is firmly bound to macromolecules like metalloenzymes and nucleic acids. Alkaline phosphatase, Aminopeptidases, Carboxypeptidases, Carbonic anhydrase, DNA-polymerases, Ethanol dehydrogenase, Glutamic dehydogenase, Insulin, Lactate dehydrogenase, Leucocytic zinc protein, Metallothionein, Proteases, Pyruvate carboxylase, Superoxide dismutase, Thermolysin.	δ-amino levulinic acid dehydratase	O^6-methylguanine-DNA-methyl transferase

4.3 LIPID PEROXIDATION

The biological membranes are composed of lipid bilayers alternated by globular proteins in a state of thermodynamic equilibrium. Their integrity and proper functioning is crucial to life processes. The oxidative reaction disrupting the integrity of biomembranes is known as lipid peroxidation. It involves transformation of unsaturated fatty acids into radicals and hydroperoxy products which, in turn, further promote free radical generation. Free radicals contain unpaired electron(s) and, therefore, are, highly reactive. Examples, include superoxide anion ($O_2^{\cdot -}$), and hydroxy radical ($\cdot OH^{\cdot}$). They cause cellular damage and pose a serious threat to normal tissue functioning. Free radicals react with non-radical species to produce another free-radical. This enables them to initiate and propagate a chain reaction. Reactive oxygen species not only destroy membrane lipids and cytosolic enzyme proteins but also cause DNA breaks and genotoxicity. There are many clinically important situations in which free radicals may play a significant role in tissue injury. These include myocardial infarction, circulatory shock, ischemia/reperfusion syndromes, rheumatoid arthritis, liver diseases, immune deficiencies, chemotherapy (bleomycin, adriamycin), chemical carcinogenesis, (benzo-a-pyrine, tobacco smoke), ionizing radiation, ageing, hyperbaric oxygen and pesticide (paraquat), heavy metal (bismuth, chromium, copper, lead, mercury, silver) toxicity, etc. Therefore, lipid peroxidation and free radical generation are inherent degradative mechanisms and inhibitors of such reactions have received considerable attention of researchers in recent years.

Anionic ligands at the cell surface, which are responsible for cation adsorption, include phosphoryl, carboxyl, sulfhydryl and hydroxyl groups of proteins and lipids in membrane and cell-wall components, e.g., peptidoglycans and associated polymers. Adsorption and distribution of metal compounds is influenced by their lipid solubility. Generally, organo-metallic compounds exhibit better lipid solubility as compared to inorganic compounds. This enables them to readily penetrate lipid membranes and elicit intracellular effects.

Under normal conditions, 95-98% of molecular oxygen consumed by cells is reduced to water by the addition of four electrons. The remaining 2-5% is reduced by univalent pathway.

$$O_2 \xrightarrow{e-} O_2^{\cdot -} \xrightarrow{e-} H_2O_2 \xrightarrow{e-} {}^{\cdot}OH \xrightarrow{e-} H_2O$$

Detoxifying enzymes, such as superoxide dismutase, catalase and peroxidases, catalyze dismutation and reduction of reactive oxygen intermediates. Ubiquinone or coenzyme Q, tocopherol and sulphydryl compounds (particularly tripeptides, cysteine, and glutathione) serve as free radical scavengers. The SH-groups present in cell membranes have strong affinity for heavy metals and therefore are the chosen target sites of damage by metallic compounds. Metals elicit their enhancing or protective effects on lipid peroxidation by interfering with mechanisms described above and as shown in Table 17.

TABLE 17 EFFECT OF SOME METALS ON LIPID PEROXIDATION AND FREE RADICALS

METAL	EFFECT/MECHANISM
Arsenic	Enhancement. Inhibition of SH-group of enzymes.
Cadmium	Enhancement. Inhibition/depletion of glutathione peroxidase, glutathione, catalase, etc.
Chromium	Enhancement. Cr (III) forms complexes with appropriate lipophilic ligands diffusing into cells with relative ease. Cr (VI) permeates into membranes. It oxidizes protective thiol groups to disulfide bridges.
Copper	Enhancement/Protection. Cu (II) and Cu (III) form complexes with appropriate lipophilic ligands and diffuse into cells with relative ease. They bind to SH-group proteins, oxidize thiol groups to disulfide bridges, form stable complexes with glutathione and reduce the availability of antioxidants. Enhancement of lipid peroxidation makes Cu (II) compounds extremely toxic. Tyrosinase, a copper containing enzyme, oxidizes catechols to a copper containing enzyme, oxidizes catechols to form orthobenzosemiquinone free radicals. Ceruloplasmin/laccase oxidizes hydroquinones and forms a molecular link between copper and iron metabolism. Electron spin resonance studies reveal that the catalytic action of ceruloplasmin involves oxidation of the electron donor to its free radical form with adverse effects.

METAL	EFFECT/MECHANISM
	Copper is a constituent of superoxide dismutase and phenol oxidase enzymes present in erythrocytes which protect biological membranes against superoxide anions.
Lead	Enhancement. Formation of stable complexes with glutathione leads to reduced antioxidant levels and stimulation of lipid peroxidation.
Manganese	Enhancement/Protection. Manganese *per se* and in the form of lectins interferes with the ultrastructure and function of membranes. It reduces the level of lipids in blood serum and protects against fat-deposition in liver.
Mercury	Enhancement. Differential solubility of inorganic and organic mercury compounds makes different cellular targets for these compounds. Binding to the SH-group of membrane proteins, inhibition of a variety of intra-cellular enzymes, particularly thiol-containing enzymes (e.g. glutathione peroxides), formation of stable complexes with glutathione and reduction in antioxidant availability account for the lipid peroxidative and cytotoxic effects of mercury.
Nickel	Enhancement. The compound nickel carbonyl, because of its lipid-solubility and volatility, forms extremely toxic vapours which cross the blood-brain barrier. These vapours cause peroxidative degradation of membrane lipids and increase their permeability to various membranes including the pulmonary alveolar membrane. Other nickel compounds also enhance lipid peroxidation, perhaps through a mechanism involving increase in tissue iron level and decrease in glutathione peroxidase activity.
Selenium	Protective. Component of glutathione peroxidase, involved in the formation and utilization of glutathione and biosynthesis of coenzyme Q, and required for removal of H_2O_2 and peroxidation products from the cells. Serum-selenium levels exhibit antioxidant properties. Selenium deficiency may render the deficient population prone to cancer induction.
Silver	Enhancement. Depletion/inhibition of glutathione peroxidase enzyme.
Vanadium	Enhancement/Protection. Inhibits ubiquinone or coenzyme-Q required for electron transport in mitochondria and scavenging of free radicals. Inhibitory effect is also elicited on acetyl coenzyme A, resulting in reduced biosynthesis of cholesterol, triglycerides and phospholipids.

METAL	EFFECT/MECHANISM
Zinc	Protection. As a constituent of superoxide dismutase, it protects against damage by superoxide anion. Moderate quantities stabilize cell membranes and protect them from free-radical injury. The mechanism of this effect may involve the lipid component of membrane.

4.4 SYSTEMIC EFFECTS

4.4.1 Liver

Arsenic

The hepatotoxic effects of arsenic in man and animals are well documented. Accidental poisoning may occur through insecticides and rodenticides. The element may also be ingested in dangerous amounts through opium procured from illegal sources and adulterated with unknown arsenic compounds to increase its claimed aphrodisiac activity. Chronic poisoning due to high arsenic content in well waters and vegetables (particularly onions and radish) has been recorded. Industrial or occupational exposure may cause various disorders in liver, e.g. cirrhosis, fibrosis, non-cirrhotic portal hypertension, choleostasis, moderate metabolic disorders, hepatic cancers and necrosis. Intake of some ayurvedic and unani arsenic preparations has been reported to cause Indian Childhood Cirrhosis (ICC) in the off springs. The post-mortem analysis of such patients showed markedly higher concentrations of hepatic arsenic. Higher hepatic arsenic content has also been reported in all types of cirrhotic diseases including alcoholic cirrhosis.

Barium

The metal has been considered as a hepatotoxin causing fatty liver. Lower, blood-barium level has been reported in choleocystitis, hepatic cirrhosis and primary carcinoma of liver.

Beryllium

Acute poisoning with soluble beryllium compounds is predominently manifested by necrosis of the medium zone of the liver. Kupffer cells reliberate the toxin to the detriment of hepatocytes.

Cadmium

The metal accumulates preferentially in the liver. It is present in negligible amounts at birth, gradually increases with age and is responsible for extensive peri-portal, inter-lobular fibrosis, biliary hyperplasia and other forms of liver damage. Cadmium inhibits both the content and activity of drug-metabolizing enzymes. Hepatic damage is one of the manifestations of acute cadmium toxicity. Measurement of hepatic and renal cadmium concentrations, using neutron activation analyses, have been suggested to assess the degree of exposure to this metal. High protein diet (particularly cysteine rich) protects the liver from cadmium-mediated damage.

Chromium

Following oral ingestion or inhalation, chromium accumulates in the liver. In human beings, chromium intoxication leads to two types of syndromes: pulmonary, and gastric. However, in both the cases, it is associated with liver damage ranging from an increase in the size of the organ to cirrhosis and organ dysfunction. In an unexposed population, the blood-chromium level tends to decrease considerably in hepatic disorders. Experimental chromium deficiency in animals results in a reduced response of isolated liver mitochondria to insulin.

Cobalt

The liver is the major site for cobalt metabolism. Both as a free metal and in complex forms like cobalamin (vitamin B_{12}) or enzymes, it is accumulated in the liver. This is presumably due to its relationship with portal circulation. Although liver vitamin B_{12} is usually believed to constitute the body stores, experimental studies in rats indicate rapid fluxes of cobalt between liver and rest of the body. During depletion states, hepatic vitamin B_{12} may not be readily available to the rest of the body.

In liver diseases (cirrhosis, viral hepatitis, malignant liver disease, etc.), the concentration of vitamin B_{12} in the liver is often reduced and that in the serum elevated. Impaired hepatic uptake in diseased states was confirmed by tracer studies. A reduction in hepatic cobalt concentrations in these diseases is common. Considerable variation has, however, been observed in its serum levels. It is likely that different types of cell injury produces different responses.

Copper

Increased copper and copper-bound protein concentrations have been reported in a number of hepatic ailments, both in adults and children. The examples include neonatal hepatitis, ICC, alcoholic hepatitis, primary biliary cirrhosis, intra- and extra-hepatic biliary atresia, haemochromatosis, fatty liver, neoplasms, Wilson's disease, etc. Wide variations in copper concentrations, reported by different workers in normal and diseased liver samples, do not permit to draw any definite conclusion in this regard. The utility of hepatic copper estimations appears to be more in cases of prolonged choleostasis, chronic biliary cirrhosis and Wilson's disease. Values of copper levels in these diseases may be upto thirty times more compared to the control levels.

Elevated serum copper levels have been reported in a wide variety of diseases, such as acute viral hepatitis, ICC, Laennec's cirrhosis, choleothiasis, chronic active hepatitis, obstructive and haemolytic jaundice, malignancy, alcoholism, alcohol withdrawl syndrome with prolonged hallucinatory state and delirium tremens.

Gold

Gold is a hepatotoxin and so its use in rheumatoid arthritis often causes liver damage. Gold thioglucose (GTG) may also cause hepatoma.

Iron

Liver, which contains 20% of the total body burden of iron, plays a central role in the metabolism and pathophysiology of liver. Lower hepatic iron concentrations are observed both in fatty and non-fatty cirrhosis, haemochromatosis and alcoholic cirrhosis with visible haemosidrin deposits. Serum-iron levels are elevated in infectious hepatitis and obstructive and non-obstructive toxic hepatitis and depressed in portal cirrhosis, kwashiorkor disease, and obstructive and haemolytic jaundice. The differentiation between toxic hepatic damage and obstructive jaundice may be based on such estimations. Besides, serum-iron, ferritin, and total iron-binding capacity (TIBC) estimations corroborate the diagnosis based on conventional liver-function tests in hepatic ailments. The ratio of serum ferritin to transaminase activity is a useful index of iron storage status. Iron is one of the major constituents of biliary calculi.

Lead

Increased hepatic lead content was reported in haemochromatosis (upto seven times the control value), Laennec's cirrhosis and Wilson's disease. Experimental lead toxicity has been shown to decrease hepatic ascorbic acid content in albino rats fed on low or high protein diet.

Manganese

It has been reported to be a hepatotoxin that causes perilobular necrosis, fatty liver and cirrhosis. High liver and serum manganese concentrations are found in acute viral hepatitis. The values return to near normal levels during the subsiding phase of the disease following treatment. In patients with ICC, manganese levels have been found elevated in hair and liver biopsy samples, depressed in nails and urine, but unchanged in serum. Alcoholics and iron-deficient subjects are more prone to the toxic effects of manganese. The metal has been detected as a major component of biliary calculi.

Mercury

Many ayurvedic and unani drugs used in India contain high concentrations of mercury. Enzyme inhibitory and hepatotoxic effects with some of such preparations have been reported. In the Minamata disease, liver accumulates substantial amounts of mercury.

Nickel

High hepatic and low serum-nickel concentrations have been observed in patients with fatty and non-fatty cirrhosis.

Selenium

This metalloid is an important constituent of Factor-3 and plays a protective role in hepatic necrosis. In animals, selenium is necessary for the prevention of several diseases which show variable response to vitamin-E treatment, for example, hepatosis dietetica and liver necrosis. The latter appears to be related to the deficiency of glutathione. Evidence for the essentiality of selenium and its human deficiency is equivocal.

Low serum-selenium levels are observed in alcoholism, hepatitis, cirrhosis including ICC, hepatic necrosis, kwashiorkor and Hodgkin's disease. The deterioration of liver function (irrespective of its etiology) leads to decrease in serum selenium levels.

Tin

Tin coating of copper/brass utensils/used for boiling/storing milk prevents copper contamination and reduces the risk of ICC.

Vanadium

Feeding of sodium metavanadate caused a significant reduction in coenzyme A content of rat liver. Children suffering from the kwashiorkor disease have been reported to have lower serum-levels than the matched controls.

Zinc

The role of this metal in the pathogenesis of liver disease is more often associated with its deficiency rather than its excess. Low serum/plasma zinc concentrations are observed in infective hepatitis, alcoholism, alcoholic cirrhosis, Laennec's cirrhosis, ICC, post-necrotic cirrhosis, chronic active hepatitis, hepatosplenomegaly, Hodgkin's disease and Wilson's disease. Hyperzincuria occurs in infective hepatitis and cirrhosis of liver. The urinary zinc levels return to normal when the disease subsides. Lower hepatic and hair zinc were reported in ICC. The metal is a minor component of biliary calculi. Human hepatoxicity with zinc has not been reported.

4.4.2 Kidney

Arsenic

Arsenic acid and some other arsenicals cause tubular damage including necrosis and have been used for the production of experimental renal toxicity in rabbits. Mean concentration of arsenic in many tissues (e.g., kidney, muscle, liver, spleen and lungs) was found to be significantly higher in uremic patients as compared with those in control subjects. Haemoglobinouria occurs in cases of toxicity manifested following the exposure of arsine and As (III) halogenides. Urinary arsenic concentration is a good index for assessing the degree of arsenic intoxication.

Cadmium

The metal is bound to metallothionein in liver and the complex so formed is transported to kidneys. Its concentration is higher in the cortex region than in the medulla. It has been observed to diffuse through the glomerular membrane into the tubular fluid and then getting completely absorbed at the proximal renal tubule. This leads to glomerular damage thus causing an enhanced excretion of high molecular-weight proteins in urine. Tubular damage may also occur alongwith glomerular damage, or even independently. This is characterized by an increased excretion of low molecular-weight proteins in urine, which is a characteristic feature of cadmium intoxication. These proteins are electrophoretically different from proteins excreted in other forms of metal poisoning. The indicator proteins in urine are β_2-microglobulin and the more stable retinol-binding protein. Renal damage always causes enhanced cadmium excretion in urine. Histological changes, however, are often absent or mild. Industrial exposure to cadmium may cause nephritis. Advanced cadmium nephropathy is often associated with the altered absorption metabolism and excretion of calcium, phosphorus, glucose and amino acids and demineralization of bones.

Chromium

Chromium salts produce renal tubular damage very similar to that observed after mercury poisoning. Rats receiving potassium dichromate (10 mg/kg s.c.) develop oligouria, proteinuria and glycosuria followed by diuresis. The histological changes include vacuolation, pycnosis, karyolysis, loss of mitochondria, fatty changes, cell rupture, cast formation, etc. Cellular regeneration with partial return of reduced enzyme activity is complete within three weeks following exposure.

Cobalt

Cobalt-induced biosynthesis of erythropoietin occurs primarily in kidneys and liver. Its chloride salt has been used for experimental induction of nephrotoxicity in rabbits.

Copper

Elevated urinary copper excretion as a non-dialysable protein complex occurs in nephrosis. Increased renal excretion of copper in nephrotic patients can be correlated to high urinary elimination of iron, zinc, and some proteins firmly bound to these metals. Excessive copper accumulation is observed in kidneys of patients suffering from Wilson's disease. This is accompanied by low serum-ceruloplasmin levels and high urinary excretion. The latter may reach 1500 µg day as against mean urinary copper excretion of 21 5.2 µg/day in normal subjects. In renal failure/uremia, serum copper levels are significantly raised.

Nephrotic effects are observed following parenteral administration of copper salts. Rats receiving cupric chloride injections (1-4 mg/kg of Cu) for 236 days were reported to show sloughing and degeneration of epithelial cells of the proximal convoluted tubules. These cells contained yellow droplets with positive histochemical reaction for copper. The other parts of nephron, however, remained unchanged.

Gold

Gold administered in the form of soluble organic or inorganic salts causes tubular damage in experimental animals. Nephrotoxicity is the most common side effect of gold therapy in patients with rheumatoid arthritis.

Lead

Renal-lead content increases with intake. It may produce acute or chronic nephropathy depending on the dose and duration of exposure. Decreased tubular reabsorption of lead results in tubular damage manifested by generalized aminoaciduria, glycosuria, hyperphosphaturia and hypophosphatemia. Prolonged exposure may lead to contracted kidneys with arteriosclerotic changes, interstitial fibrosis, glomerular atrophy, hyaline degeneration of vessels and renal failure.

Manganese

Experimental deficiency in mice is reported to cause

ultrastructural changes (integrity of cell membranes, swelling of endoplasmic reticulum, changes in mitochondria) in kidney cells. No reports of any renal disease, due to manganese deficiency or excess, have so far been observed in man.

Mercury

Mercurial preparations, such as mercaptomerin, mersalyl, mercurophylline, etc. are known to elicit diuretic effects by reducing the tubular reabsorption of water. Diuresis is accompanied by saluresis (loss of Na, C1) which is the first stage of the toxic effect of mercurial diuretics on renal tubules. In mercury toxicity, the presence of casts, albumin, leucocytes and erythrocytes in urine may be observed due to renal irritation. Organic mercury compounds, administered to pregnant females, may pass through the placenta and accumulatc in foetal organs including the kidney. Postnatal exposure through breast milk of such mothers can produce profound changes in the epithelial cells of the proximal convoluted tubules of the developing kidney. British Anti Lewisite (BAL) is useful in cases of inorganic mercury poisoning but contraindicated in methyl-mercury poisoning. Though wide variations are observed in the mercury content of human urinc, the critical urinary concentration above which mercury poisoning can be suspected has been suggested as 1-2 µg/ml.

Nickel

Dystrophic changes are observed in kidneys of animals intoxicated with nickel. Renal parenchyma, distal tubular lining of cells and mitochondria from kidneys of such animals show high nickel concentrations. Persons suffering from diffuse glomerulonephritis exhibit hypernickelemia and reduced concentration of nickel in urine. Uraemia patients may reveal low serum-nickel levels.

Selenium

Kidneys have a high concentration of selenium. Low serum selenium level has been reported in several renal diseases. The levels in infants (1 month to 1 year) with the Maple Syrup urine disease were found to be 2-8 μ g/1 as compared to the concentrations of 18-112 μ g/1 in healthy infants of the same age-group. Nephrosis is one of the symptoms of selenium intoxication.

Thallium

Organs and tissues of experimental animals, according to the degree of tropism of thallium isotope and vulnerability to histological, histoenzymatic and histochemical lesions twenty days following its intoxication, may be categorized as follows :

Kidneys, testes > liver > spleen > prostate > brain > fur

Kidney and testes seem to be the target organs of thallium toxicity.

Tin

A single injection of tin chloride (5.6-56.4 mg/kg s.c.) induced 20 to 30 fold increase in the haeme-oxidation activity in rat kidneys, 16 h after its administration. The effect was discernible at the lowest dose and was found to be dose-related. High doses of both inorganic and organic tin compounds produce renal damage including haemorrhagic lesions, fatty degeneration of cortical tubular epithelium and necrosis in experimental animals. More than double of serum-tin levels were reported in uraemic patients (75.1 ±38 μg/1) as compared to those in normal subjects (37.2±13 μg/1).

Vanadium

Kidneys are a critical organ for vanadium poisoning.

Toxic exposure of vanadium causes grave dystrophic changes in the renal tubules. Such effects and compromised tubular excretion, observed in both acute and chronic intoxication, are irreversible, even after the exposure to this metal is stopped. Correlation between urinary-vanadium levels and duration of exposure/ employment has been reported among the vanadium-based industry workers.

Zinc

Decreased activity of several zinc-containing and zinc dependent enzymes has been observed in kidneys of zinc-deficient animals. Low plasma/serum zinc levels are observed in patients of uremia. In nephrosis, the urinary excretion of zinc is well above normal. Urinary-zinc concentrations in six patients with albuminuria were found to be 1.0-3.8 mg/day as against only 0.1-0.7 mg/day in controls. Such heavy losses of zinc may be attributed to the inability of a nephrotic kidney to provide effective barrier for protein-bound zinc in plasma. Porphyrinuria is frequently observed in zinc-production workers.

4.4.3 Nervous System

Aluminium

Chronic focal seizures have been produced in experimental monkeys using aluminum hydroxide. Serum-aluminum concentrations of more than 300 μg/1 produce neurological lesions. Such levels have been shown to be attained by toxic exposures during medication. The exposure to compounds of this metal have given rise to encephalopathy in isolated cases in industry and on epidemic proportions in other population groups. Dialysis dementia has been observed by the usage of water containing 50 μg/1 aluminum from natural sources or when the metal was added to precipitate particulate matter in domestic water.

Arsenic

Peripheral neuropathy (with mixed motor and sensory features) may result from arsenic intoxication during accidental or homicidal ingestion of insecticides.

Cadmium

Nervous symptoms of cadmium toxicity include dizziness, headache, cramps, and loss of consciousness. Learning disability in children is associated with elevated hair cadmium concentrations.

Cobalt

As cyanocobalamin (vitamin B_{12}), it is essential for the functioning of all cells including the nerve cells. Experimental focal epilepsy can be produced by cobalt gelatin in rats. Accidental intoxication from excessive consumption of beer, in which cobalt is added to stabilize froth, may cause damage to the eighth cranial nerve, leading to deafness, coma and death. Diminished brain-cobalt values have been reported in cases of alcohol abuse. Patients with endogenous psychosis, on the other hand, exhibited highly significant elevation of cobalt levels in the caudate nucleus.

Copper

Wilson's disease, a rare autososomal recessively-inherited abnormality in hepatic copper excretion, results in toxic accmulation of copper in many organs including the brain. Neuropsychiatric manifestations of this disease include spasticity, rigidity, tremours, wild choreiform, movements, behaviour and personality changes, aggressive outbursts, etc. Significantly high serum and low CSF levels of copper were reported in patients with schizophrenia. Higher plasma copper concentrations were also seen in mentally retarded patients practising pica.*

---

*A depraved appetite or hunger for substances not fit for food, e.g., earth, coal, chalk pieces, etc.

Gold

Toxic symptoms of gold therapy may include peripheral neuritis and encephalitis with EEG-abnormalities.

Iron

Iron occurs as ferritin in brain and nerve cells. The salts of this metal have been used extensively for inducting experimental epilepsy in rats and cats. Injection of an aqueous solution of ferric chloride into the isocortex induced recurrent epileptiform discharges with focal brain oedema, cavitary necrosis and gliosis. A significant decrease in haemoglobin, hematocrit, plasma iron, total iron-binding capacity and plasma ferritin was observed in mental retardation. Post-menopausal Parkinsonism has been related to brain iron overload. Elevated iron concentrations were observed in the caudate nucleus of patients with endogenous psychosis. Deficiency of this element may also be manifested as CNS symptoms, e.g. headache, fatigue, irritability, restless legs (Ekbom's syndrome), etc.

Lead

The neurological manifestations of lead toxicity include headache, insomnia, dizziness, psychosis, encephalopathy with predominant motor features (wrist and foot drop, mania), coma and convulsions. The neuropathy improves on cessation of lead-intake. Mentally retarded children reportedly show elevated hair-lead concentrations. No correlation was observed between CSF lead levels and dementing illness.

Manganese

Rats receiving manganese in drinking water (1 mg/ml daily for 14-30 days) exhibited hyperactivity, reduction in conditioned avoidance response (CAR) and increase in the fighting score. Increased turnover of striatal dopamine and norepinephrine were suggested as factors responsible for these behavioural changes. The clinical features of prolonged occupational exposures to manganese in mine and mill workers include akinesia, rigidity or mania, sometimes progressing to Frank psychosis. Evidence of Parkinsonism may appear after many years of exposure. It has been attributed to disturbances of dopamine, GABA melanin and serotonin synthesis in the basal ganglia of brain. Pathological changes may be seen in the brain cortex, sub-cortical nuclei and spinal cord. Mental symptoms recover following cessation of exposure, but extra-pyramidal abnormalities are usually long-lasting. Its deficiency is manifested in skeletal abnormalities and postural defects (neonatal ataxia) without the development of histological lesions in brain or spinal cord. Elevated hair manganese level has been reported in mentally retarded children. A lower tissue-manganese level has been found in some epileptics, who received treatment for epilepsy, with frequent seizures. Reduced manganese availability at the neuronal level where Mn^{2+} stabilizes membrane excitability may affect epileptogenic lesions and increase the likelihood of seizure activity. No relationship could be found between the dementing disease and CSF levels of manganese.

Mercury

Exposure to both organic and inorganic mercury compounds may lead to neurological damage and grave genetic defects. Alkyl derivatives are more toxic than other forms. The damage caused by alkyl mercury compounds to the nervous system is often irreversible. Minamata disease is a classical example of mercury exposure, where manifestations are characterized by symptoms of fatigue, loss of memory and concentration, paraesthesia, ataxia, tremours, dysarthria, constriction of visual field, cortical blindnesss, psychosis, etc. Principal damage is caused to the granular layer in the cerebellum and visual cortex with neuronal loss and glial proliferation. In a study on 53 patients with positive methyl mercury poisoning, the frequency (%) of neurological signs and symptoms was as follows: cerebellar disturbances (95), sensory impairment (65), visual disturbance (60), speech disturbance (75), mental abnormality (50), pyramidal tract lesions (38), hearing defects (19), involuntary movements (31), cranial nerves other than optic (19), autonomic disorders (19), and encephalopathy (32).

Nickel

Marked hypothermia and diminution of metabolic rate occurs in mice after an i.p. injection of $NiCl_2$. Cerebral haemorrhages are observed in fatal cases of nickel-carbonyl poisoning.

Selenium

Selenium intoxication causes polyneuritis with symptoms of 'pins and needles' sensation, acroparaesthesia, pain in the extremities, hyperflexia of tendons, numbness, motor disturbances, hemiplegia, paralysis, convulsions and death. Patients of alcohol abuse reportedly have variable selenium concentrations in relation to specific brain functions. The values were partly diminished (cortex and muscles of the limbic systems, autonomic system, coordination of movements, vision nuclei), in the normal range (audition, voluntary movements), or elevated (tracts sensation) with alterations located mainly in the cerebral nuclei. In patients with endogenous psychosis, selenium concentrations in the caudate nucleus were significantly elevated. Brain-selenium levels were not altered in cases of schizophrenia. Mental retardation and neuronal ceroid lipofuschsinosis (NCL) are associated with low blood selenium values.

Thallium

Accidental poisoning from thallium-containing pesticides, rodenticides or depilatory agents, may cause neuropathy of mixed (motor and sensory) type with pain, paraesthesia, behavioural disorders, choreform movements, vision defects and CNS damage.

Tin

Organotin compounds have given rise to encephalopathy in isolated cases among workers in industry and on a larger scale in other population groups.

Vanadium

High dietary intake of vanadium may interfere with the sodium pump in susceptible individuals and may induce manic-depressive symptoms.

Zinc

Zinc is one of the metals found in very high concentration in the brain. It is also found in high concentrations in the choroid layer of the retina, which is an extension of the brain. The metal plays an important role in exonal and synaptic transmission and is necessary for nucleic acid metabolism, brain growth and phosphorylation. Lack of zinc has been suggested in impaired DNA, RNA and protein syntheses during brain developement. Deficiency of Zn during pregnancy and lactation has been linked to many congenital abnormalities of the nervous system in offsprings. Low learning ability, apathy, lethargy and mental retardation in children is associated with low zinc levels. Apathy, depression, anorexia, mood fluctuations and psychotic behaviour may be seen in zinc-deficient adults.

An analysis of the brain of patients with endogenous psychosis revealed significantly higher zinc concentrations in the caudate nucleus. Correlation between CSF zinc and amino-acid concentrations has been reported. Patients with damaged blood-brain barrier, subarachnoid haemorrhage, malignant tumours and high CSF protein levels exhibited higher CSF zinc concentrations while ex-heroin addicts showed low CSF-zinc levels. Untreated epilepsy patients have low serum zinc levels while no changes have been observed in schizophrenics.

4.4.4 Cardiovascular System

Arsenic

Arsenic is a capillary poison and affects tissues rich in oxidative enzyme systems, resulting in hypovolemia, shock and circulatory failure. Myocarditis, ECG changes and perivascular disturbances (e.g. Renaud's syndrome, mesenteric thrombosis and black-foot disease) may result from exposure to arsenic compounds. Arsine and As (III) halogenides are particularly toxic. They are taken up by the erythrocytes and cause their haemolysis and haemoglobinuria. Its chronic exposure may lead to bone-marrow depression, anaemia, leukopenia and basophilic stipling.

Cadmium

Highly significant differences in cadmium concentrations of normal and hypertensive subjects have been reported, with considerably elevated levels in the latter. In a study of 28 American cities, death rates from all types of heart disease (except rheumatic type) were significantly correlated to the cadmium content of urban air. Experimental induction of cadmium-induced hypertension has also been reported in rats, rabbits and dogs. The mechanism of cadmium-mediated hypertension is not clear. It possibily affects vascular resistance, sodium retention and renin-angiotensin activity. Cadmium-hypertensive rabbits, show high cadmium concentration in the heart tissue. The hypertensive effect is produced at doses insufficient to induce other signs of cadmium toxicity. At toxic levels, histological changes are observed in all vital organs including the heart. A reversible decrease in haemoglobin content is seen in the blood of exposed persons. This is attributed to its inhibitory effect on iron absorption from diet and the consequent reduction of iron supply to the bone marrow.

Chromium

Chromium has been shown to have a role in cholesterol homeostasis. Rats fed chromium (1-5 mg/ml) in drinking water or commercial chow showed depression in serum cholesterol, stainable lipids and fluorescent material in aortas and enhanced fatty acid synthesis in liver. Patients with atherosclerosis excreted lower amounts of chromium in urine.

Copper

The metal is found in moderate amounts in the heart muscle and forms an integral parts of the DNA/RNA molecule and that of ceruloplasmin and erythrocuperin in blood. An imbalance of copper and zinc metabolism probably contributes to the increased risk of coronary heart diseases (CHD). Breast milk (with optimum ratio of these two metals) is considered superior to cow milk and has been associated with lower incidence of CHD. Significantly higher serum/plasma copper levels have been reported in atherosclerosis, rheumatic heart failure and congestive heart failures. The cardiovascular symptoms of copper toxicity include tachycardia, hypertension, haemolysis, haematuria, etc. Haemodialysis may introduce sufficient copper into a patients circulation to cause haemolysis.

Iron

Iron is vital to life as an integral part of haemoglobin and the oxygen transport system. Excessive accumulation may be toxic to cells as it catalyzes the production of the hydroxyl radical. While deficiency is known to cause anaemia, excess may lead to haemochromatosis and platelet aggregation. Significant changes in serum-iron during the first seven days are a characteristic feature of myocardial infarction.

Lead

This is a highly toxic metal. It interacts with several essential elements, particularly calcium and iron and inhibits nearly all steps in haeme synthesis with primary effect on immature erythrocytes. Erythrocyte δ-aminolevulinic acid dehydratase (ALA-D) activity is the most sensitive indicator of lead toxicity. Association between excessive lead exposure and hypertension has been extensively studied. These studies suggest a link between moderately elevated blood-lead levels (>30μg/1) and B.P. increases. The association appears to be stronger for males aged 40-50 years and more for systolic than for diastolic B.P. The biochemical and physiological mechanisms by which lead influences B.P. have not been fully elucidated. However, this increase may be attributed to changes in the vascular reactivity and sympathetic tone, possibily through alteration in the intracellular calcium levels.

Manganese

The metal appears to have a lipotropic action. It is involved in cholesterol and fatty acid synthesis. The role of manganese in lipogenesis is confirmed by several studies on experimental animals. The mechanisms involved include its role as a cofactor in steroid biosynthesis, enzymatic stimulation (enzymes involved in glycolysis and pentose phosphate pathway) and through an effect on insulin metabolism. The metabolic site in cholesterol synthesis, at which manganese is required, is apparently in the formation of farnesyl pyrophosphate from the condensation of geranyl-and isopentenyl pyrophosphate. The enzymes catalyzing this reaction (farensyl pyrophosphatase synthase) are manganese-activated. In a study on a human volunteer deficienct of this metal (and vitamin K), the levels of serum-cholesterol, phospholipids and triglycerides exhibited a marked decrease. Following acute coronary occulsion, blood-manganese levels sharply elevate to about twice the normal values which may persist for several weeks. Blood manganese estimations are stated to be a better index for myocardial infarction than enzyme estimations. These aspects, however, need further validation.

Mercury

Mercurial diuretics are used principally in the treatment of oedema due to congestive heart failure. Cases of sudden death from i.v. administration of these diuretics are believed to be due to ventricular arrhythmias. Atreo-ventricular dissociation, extrasystoles and ventricular tachycardia precede fibrillation. Cardiovascular collapse with mercurial diuretics is more common in patients with nephrosis. Animal and human studies suggest a rise in B.P. following exposure to mercury in drinking water or food. The presence of mercury and selenium in equimolar ratio in marine food, with no apparent deleterious effects, indicates their auto-protective effects in nature.

Nickel

Nickeloplasmin, a nickel-containing macroglobulin, is present in rabbit and human serum. There is no clear indication of its physiological significance or function. An increase in serum-nickel has been reported in cases of acute stroke and myocardial infarction. Patients with acute myocardial ischemia without infarctions, acute trauma with fractured bones, acute delirium tremens and muscular dystrophy, exhibited normal serum-nickel concentrations. Therefore, monitoring of changes in serum nickel during the first seven days may have a diagnostic potential in myocardial infarction. Toxicity due to nickel may be manifested in cardiopulmonary insufficiency, electrocardiographic abnormalities and dystrophic changes in the cardiac muscle. Chronic intoxication with Ni carbonyl results in phased alteration of the blood-picture. During the first year of contact, stimulation of erythropoiesis and after five years of exposure a tendency to leucopenia is noted. Reductions in the haemoglobin content and erythrocyte count have also been observed in nickel toxicity.

Selenium

Autopsy samples of the kidney cortex of patients dying of cardiovascular diseases show significantly low concentrations of selenium. In a limited study, the lowest death rates from cardiovascular diseases were associated with highest selenium concentrations in the heart muscle samples from Philippines and vice versa in USA. However, when data from five countries (Philippines, Israel, USA, Greece and erst while Czechoslovakia) was scrutinized, a clear picture did not emerge. Since, platelets are considered crucial for triggering thrombosis, the high concentration (782 ± 127 ng/g fresh weight) of selenium reported in these cells, should be considered significant. Patients with thrombasthenia show marked decline in the selenoenzyme, glutathione peroxidase. Deficient states lead to an increased sensitivity to oxidative damage of membranes and oxidant sensitive proteins like haemoglobin. These aspects need to be investigated further. Symptoms of chronic selenosis include anaemia, nervous heart-beat and complaints of heartburn. Atrophic changes may be found in an autopsy sample of heart in cases of selenium intoxication.

Silver

Addition of silver at a concentration of 900 ppm (as nitrate or acetate) in diet has been shown to depress the packed cell volume and induce cardiac enlargement in turkey poults and chicks. The metal may influence cardiac function indirectly by accentuating copper deficiency, increasing selenium requirement and interacting with other elements such as zinc.

Tin

Tin compounds interfere with porphyrin biosynthesis, enhance haeme breakdown and cause reduction in haematocrit and haemoglobin levels. Anaemia associated with tin toxicity has been attributed to the induction of haeme oxygenase activity and the impairment of the cytochrome P 450 function.

Vanadium

Though there were conflicting reports in literature,

several studies indicate that vanadium compounds depress the biosynthesis of cholesterol, triglycerides and phospholipids through an inhibitory effect on coenzyme A. The latter may also reduce mitochondrial coenzyme Q. Vanadium may thus influence not only the occurrence of atherosclerosis but may also affect cardiac function through intrinsic bioenergetic mechanisms. The potential usefulness of di-ammonium oxytartrate vanadate as a hypocholesterolemic agent has been reported.

Zinc

Hair and plasma zinc concentrations are usually subnormal in atherosclerosis and myocardial infarction (MI). Highly significant differences are observed in plasma-zinc levels in MI; minimum values 75 μg/ml (control: 84.6 μg/ml) with a fall in the first three days could be taken as a reliable diagnostic test for this disease. Some reports suggest the utility of zinc therapy in cases of atherosclerosis.

4.4.5 Respiratory System

Arsenic

Acute, subacute or chronic exposure to arsenic (particularly through inhalation) may cause inflammation of the mucous membranes of nose, larynx and respiratory passages manifested by coryza, tracheobronchitis, cough and perforation of the nasal septum. Increased incidence of respiratory cancers has been reported among occupationally exposed workers of smelting units of arsenic-based industries and pesticides/spray arsenicals.

Beryllium

Beryllium exposure to workers associated with atomic energy, ceremic and fluorescent lamp industries may cause bronchitis, pneumonitis, decreased macrophage phagocytic capacity, alveolar interstitial oedema, fine miliary nodulation in the lungs, fibrosis, pneumosclerosis and lung cancer. Both acute and chronic beryllium intoxication have been reported among occupationally exposed subjects.

Cadmium

The exposure to cadmium in industries may occur from inhalation of cadmium dusts or fumes. Heated metal volatilizes and burns to form CdO fumes. It is reported that a concentration as low as 0.1 mg Cd/m^3 in air may cause toxic manifestations and levels of 1 mg Cd/m^3 and above may be fatal. The symptoms of Cd poisoning which are often mistaken for "flu", include dry sore throat, cough, severe chest pain and marked dyspnoea. The pathological changes include interstitial pneumonitis and pulmonary oedema. The death following heavy exposure to cadmium is due to generalized lung damage. In chronic cases, rhinitis, dryness and ulceration of nasal passages, loss of olfactory function, bronchitis, cough and emphysema may be observed.

Chromium

Workers exposed to chromate dust for extended periods suffer from perforated nose walls, loss of smelling capacity and ulcers. Asthmatic bronchitis has been reported as an allergic reaction to Cr (VI) compounds. A sensitized person may have asthmatic attack 4-8 h following inhalation of chromate dust or even after dermal exposure to Cr (VI) impurities in cement. The industrial exposure through inhalation may also lead to lung cancer.

Cobalt

The occupational exposure to cobalt-containing dusts may lead to progressive pulmonary fibrosis and other forms of chronic lung damage. Such patients exhibit symptoms of dyspnoea and olfactory dysfunction and may also suffer from hard-metal fibrosis of the lung.

Copper

The inhalation of dusts, fumes or mists of copper compounds causes congestion of nasal and mucous membranes with occasional perforation of the nasal septum. Its chronic exposure may lead to pulmonary copper deposition and fibrosis. Lungs of vineyard spray-workers have been observed to be blue, suggesting excess of copper and may show granulomas and malignant tumours. Abnormally high serum-copper levels are observed in chronic infectious diseases (e.g., tuberculosis) and bronchial carcinoma.

Iron

Shortness of breath is a common symptom of iron deficiency. Excess iron catalyzes the production of $^{\bullet}OH$ and $O_2^{\bullet -}$ radicals. These radicals may induce peroxidation of lysosomal membranes, depolymerize hyaluronate, release serotonin and cause endothelial and lung damage.

Manganese

Manganese-induced pneumonia has been reported in persons inhaling a high concentration of this metal while working in manganese mills, mines, smelting works, battery factories and welding jobs involving the use of electrodes containing manganese. The pneumonia so induced cannot be differentiated clinically from non-occupational broncho- and lobar pneumonia but the therapeutic effect of antibiotics is usually reduced where its eteology is manganese.

Mercury

Acute inhalation of elemental mercury causes cough, chest pain, dyspnoea leading to bronchitis and pneumonitis. Chronic exposure to organic mercury compounds may be manifested, besides other symptoms, by nasal catarrh and epistaxis.

Nickel

Lung is the primary target organ for nickel toxicity. The signs and symptoms include dyspnoea, cough, intense chest pain, cyanosis and bronchoalveolar hyperplasia. Industrial exposure may cause lung cancer. Employees of electroplating and electrolytic refinery works, who inhale vapours of water soluble nickel salts, develop chronic respiratory diseases, such as asthma, bronchitis, rhinitis, sinusitis, nasal polyps, pneumoconiasis and perforation of the nasal septum.

Platinum

The platinosis syndrome exhibits irritation of the respiratory passages with asthma-like symptoms.

Selenium

No major effects are observed on the respiratory system except irritation of mucous membranes. A garlic-like odour has been reported with excess ingestion or inhalation of selenium.

Silver

Chronic silver poisoning (argyria) manifests itself as blue pigmentation of respiratory passages. Cases of acute intoxication, though rare, have been reported with severe injury to the respiratory centres.

Tin

A brief contact with alkyl tin compounds elicits an irritant effect on respiratory mucosa. Prolonged inhalation of tin dusts causes benign pneumoconiasis. Toxicity may occur not so much from tin as from contaminants (As, Sb, Pb, etc.) in tin compounds.

Vanadium

Air-borne vanadium (e.g. V_2O_5 in steel mills) principally affects the mucous membranes. Irritative respiratory symptoms have been reported with oxides and salts of this metal, both in man and animals.

Zinc

The inhalation of zinc oxide fumes (particle size $< 1\ \mu m$) causes metal-fume fever. Its respiratory manifestations include sore throat, cough, hoarse voice, chills, fever and sometimes, bronchospasms, pulmonary oedema and pneumonitis. These symptoms develop immediately after its exposure. Significantly lower serum/plasma zinc concentrations are observed in cases of bronchial carcinoma and tuberculosis.

4.4.6 Skin

Arsenic

Occupational exposure of workers engaged in the manufacture and spray of arsenical pesticides leads to skin pigmentation, raindrop rash, hyperkeratosis of palms and soles followed by epithelioma of multifocal origin. The black-foot disease (dry gangrene) and chronic arsenic poisoning co-occur, indicating a cause-effect relationship between the two. Incidence of squamous cell carcinoma has been reported in some parts of the world consuming water-rich in arsenic.

Beryllium

A direct contact of skin with soluble beryllium compounds causes contact-dermatitis within a few days. The repeated exposure of the metal leads to allergic eczematous reactions. If a soluble compound is rubbed into a wound, a chronic indolent ulcer develops, which does not heal until the offending material is removed.

Cadmium

Contact dermatitis occasionally develops from topical exposure to cadmium compounds.

Chromium

Occupational contact-dermatitis often occurs among workers of cement and tanning industries. Chromates may even cause chronic skin ulceration. A few reports also suggest an association between lepromatous leprosy and allergic contact-dermatitis due to the exposure of chromium in cement.

Cobalt

Sensitivity to cobalt is commonly found in association with chromium or nickel causing metal-dermatitis, particularly in the hands of women. Cobalt chloride solution (1%) is included among the European standard battery of 24 patch tests for contact dermatitis.

Copper

Copper has a special role in vitiligo or leucoderma. Though the precise etiology is not fully understood, the disease is attributed to disturbances in the synthesis of melanin, which involves tyrosinase, a Cu-containing enzyme. Both elevation and depletion in serum-copper levels in patients suffering from leucoderma have been observed. Increased activity of serum ceruloplasmin is a characteristic feature of most skin diseases, including lepromatous leprosy. Copper compounds possess antiseptic, germicidal and fungicidal properties and are used in topical skin preparations.

Gold

Toxic effects of gold therapy include diffused-slate coloured pigmentation, lichenoid dermatitis, erythema, urticaria and rashes.

Iron

Iron overload is one of the etiological factors of porphyria cutanea tarda, which is characterized by skin photosensitivity, blistering and scattering pigmentation with hypertricosis. Increased serum-iron levels have been observed in a study of 500 patients of vitiligo. Leprosy is usually associated with anaemia. Depleted serum-iron and elevated skin-iron concentrations have been reported in the reactive phase of lepromatous leprosy. However, treatment status and clinical improvement could not be related to decrease in cutaneous iron.

Lead

Lead carbonate is depilatory while its acetate salt is used as a hair dye.

Mercury

Inorganic and organic mercury compounds elicit antiseptic, parasiticidal and fungicidal properties. The alkyl and alkoxy compounds are very toxic, causing local irritation and systemic poisoning when absorbed. Hypersensitivity to skin reactions may result from the use of mercurial diuretics in some patients. The symptoms may vary from mild erythema involving the area of contact to severe generalized skin eruptions.

Nickel

Allergic contact dermatitis has been reported in sensitive female populations from hair-pins, imitation jewellery, brassier hooks and other metal fastenings on clothing. Dermatitis is also seen in the palms of hair dressers from frequent use of nickel-containing shampoos. Nickel sulphate solution (5%) is included in the European battery of standard patch tests. It is interesting to note that experimental nickel deficiency in chicken, rats and swines causes skin discolouration, dermatitis and loss of hair. These symptoms are corrected by nickel supplmentation in diet.

Platinum

Exposure to platinum salts causes allergic dermatitis and urticaria in some individuals.

Selenium

Selenium sulphide (2.5% suspension or shampoo) is beneficial in cases of seborrhoeic dermatitis, psorariform seborrhoea, seborrhoea oleosa, acne vulgaris, atopic eczema and dandruff. Because of its systemic toxicity, thorough washing of hands and finger nails is recommended to remove traces of drug following each application.

Silver

Some silver compounds (e.g. nitrate, allanotoinate, sulfadiazine, etc.) are used for local antiseptic and caustic effects. Slate-blue pigmentation of the elastic fibres of skin is a characteristic feature of chronic silver poisoning. It starts on the face, hands and finger-nails which may later spread all over the body. The pigmentation is usually permanent.

Tin

Experimental deficiency of this metal is reported to cause seborrhoea and loss of skin-tonicity and hair. This could not, however, be prevented by tin-supplementation in diet.

Zinc

Deficiency of Zn is known to cause acrodermatitis enterpathica in man and rough skin in animals. The element possesses antiseptic, astringent, styptic and healing properties and finds place in many skin ointments, lotions and dusting powders. Increased urinary excretion and decreased plasma levels are general features of infectious diseases including leprosy. In lepromatous leprosy, significantly low serum and high cutaneous zinc concentrations are observed. A redistribution of the metal between blood and tissue compartments is brought about through the release of leucocyte endogenous mediator (LEM) by a continued phagocytosis. LEM is believed to be a key factor in zinc homeostasis during infection. The practical utility of these obeservations remains elusive.

4.4.7 Fertility and Sterility

The tropicity of some heavy metals to gonads is as follows :

$$Te > Cd > Hg > Ag > Ba > Ni > Zn$$

Arsenic

In a study on the essentiality of arsenic in goats, it has been demonstrated that its deficiency results in a significant reduction in the percentage of normal births, birth weights and life expectancy of the new-born.

Cadmium

A higher cadmium content is reported in males as compared to females. This was initially attributed to usually higher cadmium exposure by smoking and at places of work. However, studies in animal model systems revealed a real sex dimorphism. Further, a high estrogen concentration enhances cadmium excretion in faeces. The metal is known to affect sex hormones and reproductive function. These include ovicidal and anti-implantation effects in females and testicular damage with reduction in sperm count in males. A single intra-testicular injection of cadmium chloride (1-2.5 mg/kg) in one testicle, damages the contralateral organ as well. Regeneration of testicular tissue has not been observed following cadmium-mediated injury. The metal causes marked inhibition of succino-oxidase activity and reduction in the zinc content of dorsolateral prostate (DP) of experimental animals. These properties have been used for inducing sterility in scrub animals. Co-administration of the two metals antagonizes each against the other and normal zinc levels are maintained in DP. In both fertile and infertile men, however, no significant differences were discernible in blood-cadmium levels.

Chromium

In an Indian study, chromium was found to enhance estrogen action and depress androgenic and myotrophic effects of testosterone propionate. A progressive enhancement of urinary chromium excretion and a decrease in hair-chromium levels have been observed with advancing pregnancy but no correlation could be established between the hair content of this metal and the number of pregnancies.

Cobalt

The metal was shown to inhibit myotrophic effects of testosterone propionate but it elicited no effect on estrogen activity. After a single intratesticular injection, cobalt ions reach rat prostates. Cobalt nitrate is reported to provide some protection to testes from the injurious effects of cadmium chloride.

Copper

Copper salts have been used for experimental induction of ovulation in laboratory animals. A study on the cervical mucus of 17 regularly menstruating women revealed significantly lower copper values in ovulatory and post-ovulatory phases as compared to those in the pre-ovulatory phase. Addition of copper ions to ovulatory mucus alters its rheological properties, thereby making it unfavourable for sperm penetration. This property is the basis for use of copper in intra-uterine contraceptive devices (IUCD). Copper IUCD induce several enzymatic and hormonal changes, e.g. stimulation of estrogen, catalase and monoamine oxidase activities, depression of androgenic and myotrophic effects of testosterone, inhibition of uricase, etc. Brass (Cu: 80%, Zn : 20%) has been shown to reduce the percentage of motile sperms *in vitro*. A study on 80 infertile men and 88 men of known fertility, revealed higher plasma-copper concentrations in the former. No differences were, however, observed in the copper concentrations of seminal plasma of fertile and infertile men. Some other reports indicate lower copper values in oligospermia and azoospermia. Male intravasal copper contraceptive devices have not progressed beyond experimental investigations.

Serum-copper levels exhibit a significant rise with the progress of pregnancy. After the second trimester, a slight decline is noted but the values tend to rise again towards the onset of labour. Marked elevation in serum copper has been reported at term. Additional supply of micronutrients (including Cu) during pregnancy has been advocated. Oral contraceptives induce alterations in serum copper. Infertile women were shown to have lower serum-copper concentrations as compared to control subjects. The values were somewhat lower in women with secondary rather than primary infertility.

Gold

Gold thioglucose is reported to elicit an anti-ovulatory effect in rats and mice through the inhibition of CNS-pituitary mechanisms.

Iron

Several studies report alterations in blood/serum iron (as ferritin, transferrin, haemoglobin) levels in various phases of pregnancy and lactation. Blood-iron levels remain practically constant at a reduced level from the first few weeks upto the 40th week while a continuous depression in serum-iron is observed after the second trimester. This suggests its active transport across the placenta to the foetus. Iron supplementation helps to reduce the frequency of its deficiency, which is manifested in anaemia and premature births.

The metal affects the activity of sex hormones. It has been observed that while estrogen activity is inhibited and myotrophic activity of testosterone depressed, no appreciable effect has been noted on androgenic activity. The deficiency of iron is associated with dwarfism and hypogonadism. Low iron-concentrations are a characteristic feature of the ovulatory phase of cervical mucus. Experimental studies in rats and monkeys revealed that iron salts wreak irreversible damage to the testes and accessary genital organs. The damage and consequent sterility could not be reversed by hormones or metallic salts.

Lead

Exposure to this metal has been shown to shorten gestation, cause spontaneous abortions and reduce foetal size and growth. While no conclusive reports are available regarding the effects of lead on human ovarian function and female fertility, high exposure levels in males were shown to exert adverse effects on the testes, resulting in depressed or abnormal spermatogenesis. It has also been observed that even its low level exposures may elicit negative influence in several parameters related to reproduction such as number of sperms, hormone chemistry, placental function, foetal growth and development including mental development. Nevertheless, in a study on 18 infertile and 38 men of known fertility, no significant differences were observed in blood-lead levels.

Manganese

The deficiency of this metal is associated with menstrual disturbances, impaired reproductive function, stillbirths, low birth-weights, infant mortality, infertility due to disruption of the estrus cycle and ovulation in females and loss of libido, and inhibition of sperm mortility and sterility in males. Adenylcyclase (Mn^{2+} sensitive) activity is reported to have a positive correlation with the sperm-motility index. Elevated manganese levels, observed in the cervical mucus during the post-ovulatory phase, make it unfavourable for sperm migration. An analysis of scalp and pubic hair for manganese content showed no changes during pregnancy.

Mercury

The toxic effects of mercury on the reproductive system are species-dependent, perhaps due to inter-species differences in metabolic patterns. The metal passes through the placental barrier and may cause intra-uterine death, foetal resorption or still-births. Its alkyl derivatives are particularly fetotoxic and teratogenic. These effects came to light during epidemics of Minamata Bay and Iraq. It may also induce testicular deformation in seminiferous tubules and Leydig cells.

Molybdenum

Molybdenum was shown to potentiate the androgenic and myotrophic actions of testosterone propionate but it elicited no effect on the estrogen activity. The metal is essential during the early stages of development and its supplementation in the diet of pregnant women is recommended.

Nickel

The metal has a role to play in the regulation of prolactin. Depressed reproductive performance has been observed in nickel-deficient animals. This can be corrected by nickel supplementation. Experimental studies on its toxicity exhibited increased chromatolysis and desquamation in the spermatogenic epithelium of testes, foetal resorptions and embryonic mortality in rats.

Selenium

Studies on several species of animals indicate that selenium is necessary for normal reproductive function in both sexes. Special regulatory processes seem to ensure its preferential supply to the testes and a high concentration in sperms. The element is associated with spermatogenesis in males and the gestational process in females. Pregnant women tend to conserve selenium by decreasing its urinary excretion. Mean glutathione peroxidase activity has been found to be lower in plasma and higher in the platelets of pregnant women. The foetus accumulates selenium in the liver. However, the physiological significance of these findings is not clear.

Thallium

Investigations done on cattle from polluted areas showed high concentrations of this metal in the testes. Precancerous lesions in the female genital tract of mice and a damage to the reproductive systems with thallium intoxication have been demonstrated in both animals and humans.

Vanadium

Rats fed vanadium deficient diets (<10 ppb) for 3-4 generations exhibited reduced fertility and increased pup mortality as compared to control animals fed on normal diet. Human reports in this regard are not available.

Zinc

An increased demand and decrease in serum-zinc levels is a normal feature of pregnancy and lactation. Severally depressed serum values (<50-55 μg/dl), however, indicate an inadequate zinc status. Deficiency states have been associated with foetal and maternal complications during pregnancy, including delayed labour, increased necessity of surgical interventions during delivery, low birth weights, etc. Therapeutic supplementation, however, requires a cautious approach because of reports of some cases of premature and still births following such supplementations. The antimicrobial activity of the amniotic fluid is zinc-dependent. Amniotic fluid-zinc concentrations are also reduced by 34-36 weeks of pregnancy. This is followed by a sharp rise in its level, with the peak attained around 40-42 weeks. In cases complicated with hypertension and toxaemia of pregnancy and in deliveries with congenital defects in the new born, maternal plasma-zinc values show a marked decline. A negative correlation between copper and zinc is noted during pregnancy. Alterations in this ratio, observed in obstertric abnormalities and toxaemia cases, have no predictive utility because of a wide variation and overlapping value ranges.

Both animal and human studies reveal low blood-zinc levels in infertile males. Deficiency states are associated with hypogonadism, reduced sperm-motility and viability, chronic prostato-vesiculitis, secretary dysfunction and a decline in the fertility index of semen. In a study involving experimental human deficiency, restricted dietary intake of this metal (2-7 mg/day for 24-40 weeks) caused oligozoospermia in 4/5 volunteers. Adverse effects on the testicular function, so induced, can be reversed by zinc supplementation.

4.4.8 Immune Function

Arsenic

Arsenic elicits a biphasic effect on immune function. At low doses, it is reported to cause a proliferation of T-cells and an increase in the number of antibody-forming cells in response to mitogens and heterologous antigens. At higher doses, it suppresses the immune response. Apparent immunostimulation, observed at low doses, has been attributed to a cytotoxic action of arsenic on suppressor T-cells rather than on the enhancement of the lymphocytic function.

Beryllium

Ionic beryllium affects the skin and lung tissue components. It is reported to reduce the alveolar clearance rate and macrophage phagocytic activity. It induces hypersensitivity skin-reactions and prevents wound healing. Patch-testing cannot be safely used as a predictive test for beryllium hypersensitivity as it may itself induce sensitization in human subjects. Patients with chronic beryllium toxicity have been shown to have depressed sensitivity to tuberculin.

Cadmium

The effects of cadmium on the immune function are similar to those of arsenic. Its low doses increase the antibody-forming cells, affect the proliferative response of T-cells to mitogens, enhance the response of B-cells to mitogenic challenge, and selectively inhibit suppressor cells, which results in an increased ratio of helper to suppressor cells. On the other hand, its high doses suppress DNA synthesis at an early stage of cell activation, primary and secondary immune responses against heterologous antigens, antibody titres against infections, delayed hypersensitivity skin reactions, activity of macrophages and polymorphonuclear lymphocytes, cellular phagocytic capability, and lymphokine production. Immunosuppressive effects of this metal tend to increase the susceptibility of exposed animals to infections and cancer induction. Tissue-culture studies using human lymphocytes revealed that these cells may attain upto 3000-fold intracellular concentrations of cadmium compared to that present in the surrounding medium. Though the precise reason for such a selective and high uptake is not clear, but it may have obvious immunotoxic implications in man. Tubular and glomerular lesions, in the kidneys of populations exposed to cadmium have been attributed to its immunological effects.

Chromium

Chromium is known to cause allergic skin reactions, contact dermatitis, eczemas, skin ulcers and asthmatic bronchitis. The hypersensitivity to chromium occurs because it is a hapten which conjugates to proteins. Studies on experimental animals reveal that its exposure results in a decrease in the T-dependent antibody response, B-lymphocyte proliferation, lymphokine production, phagocytic activity, and chemotaxis, and an increase in allergenicity. This indicates immunotoxic effects against humoral immunity, cellular immunity and non-specific host defences. Potassium chromate is a very common skin-sensitizer, ranking second to nickel as a cause of contact hypersensitivity. The permeability of membranes to Cr (VI) compounds allows the entry of greater amounts of the immunologically-active metal to cells as compared to Cr (III) compounds. This makes skin more prone to the effects of Cr (VI) compounds.

Cobalt

Experimental studies on cobalt in mice, guinea pigs and rabbits reveal its immunodepressive effects both in humoral immunity (T-cell dependent antibody response) and in host-resistance assays (against viral infection and tumour induction). The deficiency states are associated with atrophic changes in gastric mucosa which have been attributed to autoimmune mechanisms forming antibodies against Castel's intrinsic factor and parietal cells. Cobalt chloride is also a skin sensitizer.

Copper

Experimental deficiency of copper in rodents causes impairment of both humoral and cellular immune responses, suggesting a role of copper in the maintenance of immunological homeostasis. Exposure to this metal has also been shown to cause a reduction in T-cell dependent antibody response in mice and a decreased chemotaxis in rabbits.

Gold

The therapeutic effects of gold compounds are attributed to their anti-inflammatory and immunoregulatory activities. Skin reactions to gold jewellery are rare but occur most frequently among patients on chrysotherapy who develop sensitivity to gold drugs.

Iron

Both deficiency and overload states of iron have been linked to disturbances in immune functions. It plays a critical role against infection. Patients with severe iron-deficiency exhibit impaired cell-mediated immune responses (phytohemagglutinin-and PPD-induced lymphocyte transformation, intracellular bactericidal capability) with no effect on phagocytosis and humoral activity. Iron overload, on the other hand, impairs phagocytosis and causes 40-50% increase in infection episodes. Parenteral administration of iron in infected animals is reported to promote bacterial multiplication and development of lethal infections.

Lead

The adverse effects of lead on the immune system are well documented. Studies in mice and rats revealed that exposures to this metal may decrease: the weight of the thymus gland, the serum immuno-globulin levels, T-cell dependent antibody response, B-cell rosetting, delayed-type hypersensitivity to sheep RBCs, phagocytosis response in host-resistance assays; and, may increase: the concanvalin A activity, passive cutaneous anaphylaxis to ovalbumin, and susceptibility to endotoxins and infectious agents. Lead-induced immuno-suppressive effects are discernible at such low exposure levels which are not associated with any apparent signs of lead toxicity. Besides, these effects are consistent and, therefore, are of value in biomonitoring human exposure to this metal. Though the effects are more pronounced in humoral immune response, cell-mediated immunity and reticuloendothelial functions may also be affected, which account for the decreased host resistance. A significant negative correlation has been reported between blood-lead concentrations and the serum levels of complement C_3 and IgG in workers exposed to this metal. Marked differences in immunotoxicity have been observed with different salts of lead. These effects also depend on the route and duration of exposure.

Manganese

Experimental studies in mice and guinea pigs showed that the administration of manganese compounds decreases the T-cell dependent antibody response, lymphokine production and host resistance to *Klebsiella pneumonae* and influenza virus infections, while it increases the natural killer-cell activity.

Mercury

Mercury compounds exhibit toxicity to various components of the immune system. Studies in experimental animals showed a decrease in the weights of thymus and spleen, in T cell-dependent antibody response, in T-lymphocyte proliferation and in concanavalin-A activity; and, an increase in B-lymphocyte proliferation and positive response to anti-nuclear autoantibodies and contact-sensitization tests. Both inorganic and organic mercury compounds enhance the mortality of animals attacked by viruses. Methyl mercury impairs both the primary and secondary immune responses. Some studies, however, reveal stimulatory effects of methyl mercury on mixed lymphocyte cultures. Immunomodulatory nephrotoxicity is believed to be immunologically mediated.

Nickel

Nickel is a potent skin sensitizer. The prevalence of nickel sensitivity is reported to be about 8-14% for women and 1% for men. Immunotoxicity and allergic reactions due to this metal are frequently observed following external exposures to its alloys (jewellery, coinage) or following oral intake or inhalation. Animal studies indicate that pulmonary alveolar macrophages are the primary targets of the immunotoxic effects of nickel. Its acute or chronic exposures to mice, guinea pigs and rabbits have been demonstrated to cause suppression of : the macrophage function, antibody response to T-cell-dependent antigens, response of lymphocytes to T-and B-cell mitogens, delayed hypersensitivity reactions and systemic natural killer cell function. Intra-tracheal nickel administration, however, enhances natural killer-cell activity in bronchial lavage cells of monkeys. In this study, the phagocytic activity also decreased with no effect on the number of primary antibody-forming cells. The clinical relevance of these findings remains to be elucidated.

Selenium

Unlike many other trace elements, selenium potentiates the immune response. An increase in T cell-dependent antibody response in man, enhanced activity of natural killer cells in Sprague-Dawley rats and decreased lymphokine production in guinea pigs following exposure to this metalloid have been recorded. Nutritional supplementation to selenium may protect animals against spontaneous and transplantable tumour induction. Deficiency of selenium has been found in cases of acquired immuno-deficiency syndrome (AIDS).

Tin

Immunosuppressive effects have been observed with triphenyltin acetate, dioctytltin dichloride, dibutyltin dichloride and tributyltin chloride. Water-soluble dimethyl and triethyl tin compounds exhibit no such effects. Studies in rodents reveal its depressant effect on weights of lymphoid organs (thymus, spleen, popliteal lymph node), T cell-dependent antibody response, B-lymphocyte proliferation, phagocytosis, concanavalin-A activity, white-blood cell count and serum immunoglobulin levels. In delayed hypersensitivity reaction tests, varying responses have been noted with different compounds of tin. For example, dimethyl tin chloride potentiates the response while triphenyl tin hydroxide suppresses it.

Vanadium

Vanadium (at a dose level between 5-10 μg/ml for 20 h) has been shown to inhibit phagocytosis in rabbits as evidenced by 45-85% decrease in polystyrene particle uptake by aleveolar macrophages *in vitro*. Altered integrity of alveolar macrophages is believed to be caused by a higher incidence of mortality from bronchitis and pneumonia at places with high concentrations of vanadium in the ambient air.

Zinc

Zinc plays an important role both in the development and maintenance of the immune system. The immune functions are affected by its decificney as well as by its excess. Deficiency states have been reported to impair cellular and humoral immunity in animals and man. Zinc deprivation influences autoimmunity in mice. The exposure to zinc causes an increase in bone-marrow precursor cells and decrease in B-lymphocyte proliferation and T-dependent antibody response. Therapy with this metal in the aged (>70 years) was found to increase the total lymphocyte counts, the serum immunoglobulin levels and the delayed-type hypersensitivity to PPD.

The metal is moderately mitogenic and impairs phyto haemagglutinin-induced lymphocyte proliferation. Its large doses may be immunotoxic. Its toxic exposure has been reported to inhibit phagocytic capacity of peritoneal macrophages and susceptibility to infection in mice. Human studies also indicated that large doses reduce : the lymphocyte transformation response to T-cell migoten PHA, chemotactic migration, and ingestion of bacteria by neutrophils. The most widely accepted factor for the metal-fume fever is an immune complex reaction to the inhaled zinc oxide fumes. The mechanisms whereby large doses of zinc alter immunocompetence include changes in the serum lipid profiles and the structural lipid composition of cellular membranes.

4.5 DISEASES

4.5.1 Deficiency Diseases

Arsenic

Its deficiency in experimental animals such as rats, chicks, minipigs and goats has been shown to cause growth retardation, rough-fur formation, enhanced osmotic fragility of erythrocytes, increased incidence of abortions, reduced milk production and high mortality with myocardial damage during lactation. However, any particular human deficiency-disease related to arsenic has not been reported.

Chromium

Deficiency of chromium may lead to diabetes, atherosclerosis, impaired growth and cataract formation in animal model systems. Human deficiency-states have been reported in elderly subjects, specially in patients on prolonged parenteral nutrition. These are associated with elevated serum lipids, glucose intolerance, which is usually refractory to insulin but responsive to chromium therapy, weight-loss and peripheral neuropathy. Decline in tissue-chromium levels with increasing age indicated a probable role of this metal in aging.

Cobalt

Deficiency of the cobalt-containing vitamin B_{12} causes pernicious anaemia. Such states, though rare, may be seen in strict vegetarians. The signs and symptoms include weakness, fatigue, sore tongue, paraesthesia in hands and feet, low erythrocyte counts (< 2 million per mm^3), enlarged blood-cell volume, elevated corpuscular haemoglobin concentration, marked hyperplasia of the rubriblasts, and neural manifestations.

Copper

Copper-protein deficiency in animals has been reported to cause anaemia, hypercholesterolemia, hyperuricemia, glucose intolerance, abnormalities of cardiac function such as aortic aneurysms, cardiac failure, ruptured heart due to defect in formation of aortic elastin, sudden death, bone disorders, depressed growth, and abnormal wool.

Menke's kinky hair syndrome or trichopoliodystrophy is a sex-linked genetic disorder caused by a defect in copper absorption and utilization. Low levels of copper in this disease lead to progressive cerebral degeneration, muscle hypotonous, seizures and abnormal hair. The disease is usually fatal in man. Other forms of human deficiency are rare and easily treatable by copper supplementation.

Iron

Iron-deficiency anaemia is prevalent on a massive scale in malnourished infants and women in India. Besides dietary intake and malabsorption syndromes, other causes which may lead to human iron deficiency state include its increased demand, for example during rapid growth in infancy and pregnancy, poor stores at birth, e.g., during premature delivery and twin pregnancy, and pathological blood loss which may be due to hookworm infestation, piles, ulcers, haematurea, menorrhagia, etc. Anaemia is initially normocytic normochromic and later microcytic hypochromic with normoblastic hyperplastic bone marrow. It may or may not exhibit symptoms such as palpitation, persistent fatigue even on adequate rest, shortness of breath, pallor, blanched mucous membranes, muscular pain, dizziness, headache, etc. The recommended daily intake for iron is about 10 mg. The requirement, however, is considerably increased during puberty, menstruation, pregnancy and breast-feeding in women. Studies on rodents indicate an increased incidence of certain types of tumours and enhanced sensitivity to a number of carcinogens. Iron deficiency is also reported to facilitate absorption of other heavy metals that include cadmium and lead. This renders iron-deficient individuals more vulnerable to the toxic manifestations of these metals.

Manganese

Manganese deficiency states have been induced in a variety of laboratory and farm animals. Its deficiency leads to retarded growth, skeletal abnormalities, motor incoordination, ataxia, convulsions, menstrual disturbances, reduced fertility, still births,azoospermia, glucose intolerance, depressed blood coagulation, decreased serum cholesterol and triglyceride levels, structural changes in mitochondria, etc. The skeletal abnormalities in manganese are attributed to alterations in the activities of manganese-dependent enzymes, such as polymerase and galacto-transferase. Impairment of bone development and growth also affects the skull formation. Neonatal ataxia is believed to be caused by improper otolith development. Precise biochemical lesion(s) underlying reproductive dysfunctions have not been defined. The suggested possibility of defective steroid synthesis in deficient animals is yet to be proved with more authentic experiments.

Human deficiency has been reported in a male volunteer, with symptoms of weight-loss, reduced growth of hair and nails, pigment changes, reddening of black hair, dermatitis, hypercholesterolemia, and prothrombin deficiency. Lack of practical methods for assessing an individual's manganese status might be hindering the diagnosis of several manganese-deficient cases and their possible links with human disorders.

Molybdenum

Dietary deficiency of molybdenum has not been reported. Tungsten-induced deficiency of this metal has been reported in rats. Low molybdenum diets have been shown to cause xanthine calculi in sheep and reproductive dysfunctions such as poor conception and low foetal survival in chicks and goats. A genetic disorder associated with molybdenum-cofactor deficiency in children is manifested in neurological abnormalities, dislocation of eye lens and mental retardation. The biochemical manifestations include increased urinary excretion of sulfite, thiosulfate, S-sulfocysteine, hypoxanthine and xanthine. Studies also reveal that changes in molybdenum *per se* have not been appreciable and its supplementation has no effect on the pattern of abnormal urinary metabolite excretion. These observations suggest that molybdenum deficiency alone is not a causative factor. Tissues of such patients lack molybdopterin, a sulfur-containing organic moiety of the cofactor. Its pure deficiency state has been reported in a human patient who was fed on total parenteral nutrition. The clinical symptoms such as irritability, tachycardia, tachypnoea, night blindness and coma, in this case, were alleviated by the supplementation of molybdenum as 300μg of ammonium molybdate per day.

Nickel

Nickel-deficiency states have been described in laboratory and farm animals such as rat, minipig, chick, cow, goat, and sheep. Signs of nickel deprivation depend on various factors, viz. species, age, iron status and interaction with atleast thirteen other trace elements including Ca, Cr, Co, Cu, I, Fe, Mg, Mn, Mo, P, K, Na, and Zn.

Despite such a wide variation, the more consistent signs include growth retardation, unthriftiness (uneven hair development, rough coat), delayed estrus, increased perinatal mortality, changes in enzyme activity, particularly urease, ultrastructural abnormalities of liver, and an altered iron metabolism leading to impaired absorption of this metal and depressed haematopioesis characterized by low values for hematocrit, haemoglobin and erythrocyte count. Human nickel-deficiency syndrome has not been reported.

Selenium

A large number of diseases and aging have been linked

to low selenium status in experimental and farm animals, such as rat, guinea pig, rabbit, hamster, dog, monkey, sheep, goat, cattle, swine, poultry and in man. These include arthritis, cancer, cataract, creatinuria, cystic fibrosis, decreased immune function, diabetic retinopathy, encephalomalacias, exudative diathesis, foetal resorption, gizzard myopathy, growth retardation, hepatosis dietetica, hepatic necrosis, hypertension, infertility, ischemic heart disease, Kaschin-Beck disease, Keshan disease, loss of hair/feathers, Mulberry heart disease, multiple sclerosis, muscular degeneration, neuronal ceroid lipofuscinosis (NCL), pancreatic fibrosis, periodontal disease, protein-energy malnutrition, renal degeneration, sickle-cell anaemia, stiff-lamb disease, testicular degeneration, thyroid dysfunction, unthriftiness, Werner syndrome, etc.

The human requirements of this metalloid have been suggested to be approximately 70 µg/day. Intake levels of 50-200 µg/day are considered adequate and safe. In certain areas of China, endemic selenium deficiency (Keshan disease) occurs with low-selenium levels in staple foods where average dietary intake is 11 µg/day. The disease is fatal cardiomyopathy, with irregular pulse, enlarged indurated hearts and mitochondrial swelling. Women and children are particularly prone to this disease. Another endemic human deficiency state, also reported from China, is an osteoarthropathy, or Kaschin-Beck disease. It involves collagen breakdown and injury to joints. Most of the signs and symptoms of selenium deficiency are partially or completely ameliorated by the administration of vitamin E. The metalloid, however, has a functional role, such as in glutathione peroxidase, distinct from that of vitamin E which has only antioxidant properties.

Tin

Total deprivation of tin in diet has been shown to cause growth retardation in experimental animals. Though a human deficiency syndrome has not been reported so far, preferential bioaccumulation of tin in thymus gland (6 times higher than in other organs) and increased blood-tin levels after experimental tumour transplantation suggests its possible role in immune mechanisms. These aspects warrant further in-depth investigations.

Vanadium

Growth retardation, weight-loss, abnormal bone development, reduced fertility, increased pup mortality, enhanced sodium retention, oedema, disturbed carbohydrate and lipid metabolism and impaired renal function have been attributed to the deficiency of vanadium in chicks, rats and in some other animals. Human deficiency syndrome has not been reported. However, there is an evidence to suggest its role in some human ailments such as nutritional oedema, maniac depressive psychosis, dental caries, diabetes and atherosclerosis.

Zinc

Zinc deficiency has been studied in many animals including rat, mice, guinea pig, rabbit, dog, monkey, poultry, sheep, goat, cattle and swine. The symptoms, observed mainly in young animals, include reduced food-intake, growth depression, skin lesions, esophageal lesions, loss of hair/feathers, skeletal abnormalities, male hypogonadism, aberrant or inhibited perception to taste/smell, impaired brain functions, impaired immune function, atrophy of thymus and lymphoid tissue, decreased packed-cell volume, lymphocytopenia, difficult parturition and tetratological effects.

Human deficiency may occur in alcoholics, patients on total parentral nutrition, infections and in several conditions involving increased zinc loss from the body, e.g., burns, diabetes mellitus, nephrotic syndrome, hepatic diseases, chronic blood loss, sickle-cell anaemia, dialysis therapy, and use of diuretics and chelating agents. Experimental zinc deficiency was produced in human volunteers by oral administration of L-histidine. Its marginal deficiencies may be difficult to detect due to its non-specific clinical features. Further, owing to wide variations and an interplay of a variety of factors, determinations of zinc levels in serum/plasma/hair samples may not be very helpful in the diagnosis of deficiency states. The activities of certain zinc-binding proteins may serve as a useful confirmatory index in such cases. The earliest symptoms of zinc deficiency are anorexia, growth retardation, impaired taste and dysomia. Other manifestations include apathy, depression alopecia, delayed wound-healing, irritability, mood fluctuations, confusion, psychotic behaviour, hypogonadism, impotence, night blindness, dysarthria, diarrhoea and weight loss.

Acrodermatitis enteropathica is an autosomal recessive-genetic zinc deficiency disease. It is characterized by skin lesions (alopecia totalis, paronychia, and bullous pustular dermatitis predominantly distributed along the extremities, oral, anal and genital areas), nutritional disturbances, diarrhoea, blepharitis, conjunctivitis, depressed mood, growth failure and frequent infections. A deficiency syndrome has also been reported in Iranian males with symptoms of anaemia, dwarfism, hypogonadism, rough and dry skin and mental lethargy.

4.5.2 Diseases Related to Over Exposure

Toxicity of various metals in terms of LD_{50} in animals is reported to fall into the following sequence:

Hg> Ag> Cu> Zn> Ni> Pb> Cd> As> Cr> Sn> Fe> Mn> Al> Be> Li

Aluminum

The metal is relatively non-toxic. The fatal doses in animals are, therefore, large, viz. cats (5-10g), dogs (35-50g), and cattle (500g). Harmful effec ts following the exposure to aluminium include gastro-intestinal disturbances, neutrotoxicity, haematological changes and bone damage.

Aluminium compounds have given rise to encephalopathy in isolated cases in industry and of epidemic proportions in other population groups. Alzheimer's disease, which is characterized by pre-senile dementia, sclerosis and neurofibril degeneration, is associated with aluminium accumulation in the brain tissue. A cause-effect relationship, however, could not be proved. Current medical opinion does not link its presence as an etiological factor but believes that the metal may be a nonspecific marker of degenerating neurons.

Dialysis osteodystrophy and dialysis dementia may result from prolonged treatment with water containing more than 50 μg/litre of Al from natural sources or when aluminium sulphate is added to precipitate particulate matter in domestic water. The clinical manifestations include bone-pain and speech disorder, followed by dementia, convulsions, myoclonus and death. Elimination of aluminium from the dialysis fluid arrests encephalopathy early in the course of a neurological disease. In a study on 258 patients treated by haemodialysis, 18 patients developed dialysis dementia. All these cases appeared in patients on home dialysis and none in patients who were exclusively on hospital dialysis. An analysis of domestic water supply of the affected subjects revealed aluminium concentrations to be more than 80 μg/litre. Reverse osmosis, which removes aluminium from the dialysis water, elicited beneficial effects in 7 out of 9 patients so treated. There is no safe level of aluminium in water used to prepare the dialysate and it is absolutely essential to remove it.

Antimony

Dermal contact with a high concentration of antimony salts causes irritation of the skin and mucous membranes. Oral intake of excess amounts results in diarrhoea, vomiting, hepatic damage, metabolic disturbances and circulatory collapse. Inhalation exposure has been reported to cause pneumonitis, fibrosis, bone-marrow-damage and carcinomas. Stibine produces adverse effects on erythrocytes and the nervous system.

Arsenic

Ingestion of a large dose of a soluble arsenic compound, especially on an empty stomach, may cause death within a few hours. The fatal dose of arsenic trioxide in adult humans is 70-180 mg. Usually, swallowing of arsenic is painless but it may cause epigastric pain in some cases. Arsine also causes death in 25-30% cases within 24 h of its exposure. The immediate symptoms are severe gastrointestinal irritation, diarrhoea, vomiting, haematuria, leg cramps, shock and coma. Death may result from acute injury to the myocardium with or without brain-stem medullary failure. The surviving patients and cases with chronic exposure to arsenic, show inflammation of the conjunctiva and respiratory mucous membranes, epistaxis, cardiomyopathy, facial oedema, skin diseases such as erythematous or vesicular rashes, scaling, pigmentation, hyperkeratosis, transeverse striate leukonchia or Aldrich-Nees lines on nails, destruction of red-blood cells and bone-marrow depression (haemolysis, anaemia, leucopenia, jaundice, haemoglobinuria), muscular pain, weakness, renal damage and symptoms of peripheral neuropathy with both sensory and motor effects such as ataxia, numbness, tingling in limbs, drowsiness, poor memory, episodes of delirium and irrational speech, paralysis, degeneration of inner ear, etc. People from an area of Taiwan, on being exposed to high levels of arsenic in drinking water, suffered from a gangrenous condition of the extremities called 'black foot disease.' Arsenicals may also act as skin-contact allergens.

Beryllium

Beryllium is the most permeable and toxic element of Group II A, thus necessitating great caution in its handling and use. It has been shown to cause degenerative lesions of the parenchyma, inhibition of bone metabolism, osteoporosis, impaired haematopoiesis, dermatitis, delayed wound healing, granulomas and lung tumours in experimental animals. Inhalation of soluble compounds of beryllium, dusts or aerosols is a serious health hazard. It produces an acute pulmonary disease with symptoms of bronchitis, pneumonitis and interstitial alveolar oedema. Its chronic exposure damages alveolar membranes, inhibits phagocytosis leading to fine miliary nodulation in the lungs, fibrosis, granulomatous lesions and pneumosclerosis. Though its systemic absorption from dermal contact is limited, ionic beryllium can react with tissue components to cause an impaired immune response.

Cadmium

Acute cadmium poisoning may result from drinking water or other fluids, particularly of acidic pH, contaminated with this metal such as resulting from the leaching of solders in water pipes, vending machines or cadmium-glazed pottery. Symptoms of exposure through ingestion or inhalation include nausea, vomiting, diarrhoea, short breath, headache, fever and choking fits. Chronic effects are lung emphysema, renal dysfunction, inhibition of iron absorption, decreased haemoglobin levels, elevated blood pressure, disturbed liver function, increased excretion of calcium and phosphorus and bone demineralization. Intake of a high-protein diet has a protective effect on cadmium toxicity. Protein malnutrition, on the other hand, exacerbates its adverse effects.

A serious form of cadmium poisoning, called the Itai Itai disease, has been reported from Toyama region of Japan. It is characterized by severe glomerular and tubular renal damage with proteinuria, bone decalcification, osteomalacia, osteoporosis, vulnerability to fractures, pain in the back and legs, and an early death. Older women, particularly post-menopause, are prone to the induction of this disease. In a case study, as many as, seventy two fractures were reported. After prolonged investigations, the disease was attributed to the intake of Cd^{2+}-laden water either by drinking it or by eating rice-crops irrigated with this water which was discharged into a river by a zinc-mining company. Cadmium exposure may not be the sole etiological factor for the Itai-Itai disease. Other contributing factors may include malnutrition, in general, and vitamin D deficiency.

Chromium

The toxicity of chromium depends on the oxidation state of its compounds. Hexavalent chromium is more toxic than trivalent. Its accidental ingestion may cause acute human poisoning. Oral uptake of 0.5 g of potassium dichromate is fatal for man. The metal is corrosive and may be absorbed through skin in dangerous amounts. The symptoms of its acute toxicity include diarrhoea, gastro-intestinal bleeding, cramps and severe damage to liver and kidneys. The numerous industrial uses of chromium are the major cause of its exposure to many workers, leading to potential ill-effects, that include skin allergies, dermatitis, slow-healing ulcers, eczema, perforation of nose walls, dysomia, bronchial asthma, and lung cancer. A solitary study on a male patient reported an association of lepromatous leprosy and contact dermatitis due to exposure of chromium present in cement. Further studies to confirm or refute such links are not available.

Cobalt

Its chronic toxicity may result from a combined therapy of anaemic patients with iron and cobalt. The symptoms include anorexia, gastro-intestinal disturbances, diarrhoea, vomiting, substernal aches, skin rashes, tinnitus, paraesthesia, neurogenic deafness, damage to the optic nerve, etc. The metal causes hypoxia through inhibition of enzymes involved in cellular respiration and citric acid cycle. This leads to enhanced erythropoietin biosynthesis and polycythemia. Industrial exposure through dusts of tungston carbide (commonly bonded with cobalt powder) and hard-metal grinding (using cobalt-containing coolants) may induce allergic alveolitis, eczema, increased erythropoiesis, decreased ability of thyroid to accumulate iodine, i.e., goitre, etc. Most of these effects are temporary and pass off following 1-2 months of cessation of exposure.

Fatal cobalt intoxication has been reported in heavy beer drinkers with symptoms of congestive heart failure, thyroid dysfunction, damage to the eighth cranial nerve, deafness and coma. This was attributed to excessive consumption (12 litres or more with cobalt intake of 6-8 mg per day) of certain brands of beer, to which cobalt sulphate (1 mg/litre) has been added to stabilize froth. It is interesting to note that intake of cobalt in much larger quantities in other forms (e.g. in anaemia therapy) was not lethal. These observations suggest involvement of other factors in cobalt toxicity. The examples include enhancement with alcohol, reduction with concurrent administration of iron, destruction of thiamine by sulfur dioxide contained in bottled beer, etc.

Copper

Ruminants are particularly prone to the toxic effects of copper. Oral intake of excess amounts of copper causes gastroenteritis, nausea, vomiting, anorexia, dyspnoea, muscular and renal damage, haemochromatosis, and fatal hepatitis or haemolytic anaemia. Acute human toxicity occurs only when large doses of soluble salts are ingested accidently or with suicidal intent. It may also result from haemodialysis or topical application of copper preparations over extensively burned areas of the skin. The symptoms are similar to those observed in cattle. Exposure to dusts, fumes and mists, in addition to producing the abovementioned symptoms, also causes congestion of respiratory membranes and perforation of the nasal septum.

Wilson's disease is a rare autosomal recessively-inherited abnormality, in which toxic amounts of copper accumulate in the liver and central nervous system. The patients, however, exhibit decreased plasma ceruloplasmin levels. The symptoms of Wilson's disease include hepatic dysfunction, haemochromatosis, yellow atrophy of liver, cirrhosis, anaemia, spasticity, rigidity, tumours, wide choreform movements, drooling, dysphagia, dysarthria, bahvaioural and personality changes including aggressive outbursts, etc. If untreated, the disease terminates in invalidity and death.

The neurological manifestations of disease may not be observed in many cases. The simultaneous presence of less than 20 mg/dl of ceruloplasmin and more than 250 μg/g of dry matter of copper in liver or detection of Kayser-Fleishcher rings in ocular cornea are considered diagnostic features of the disease.

Links between contamination of milk and drinking water by copper due to boiling/storage in brass/copper utensils and Indian Childhood Cirrhosis (ICC) have been reported by many investigators. The copper theory of the etiology of ICC, however, leaves many questions unsolved. Probably a genetic defect exists in such cases. Tin coating on brass cooking utensils prevents copper contamination.

'Excess tissue copper can cause schizophrenia' is an old concept that has never been proved or refuted with authentic experiments. The patients show high serum and low C.S.F. levels of copper. This is consistent with the hypothesis of reduced activity of copper-dependent enzymes in schizophrenia.

Copper salts have also been associated with the 'vineyard sprayer's disease,' which affects both lung and liver. It is observed among grape garden sprayers using the Bordeux mixture (which contains copper sulphate and lime). Its prolonged exposure causes hepatosplenomegaly, granulomas and portal stenosis. Periodic estimations of the serum alkaline phosphatase level, which is usually elevated, have been recommended for identifying workers prone to hepatic damage from such an occupational exposure.

Gold

Gold toxicity may occur as a side-effect of gold therapy in patients of rheumatoid arthritis with symptoms of mucocutaneous reactions, skin rash, oral ulceration, renal and hepatic damage, peripheral neuritis, encephalitis with EEG abnormalities, suppression of cells in the bone-marrow (red and white blood cells, and platelets) thrombocytopenia, and aplastic anaemia, which may sometimes be fatal. Skin reactions to gold jewellery, though rare, may be observed among some chrysotherapy patients.

Iron

Acute iron toxicity may be the result of an overdose of medicinal iron. Its symptoms include haemorrhagic gastritis, intravascular clotting, acidosis and fatal shock. Iron overload has been associated with haemochromatosis, a genetic disorder of iron metabolism characterized by a brown discoloration of the skin and other organs due the destruction of red blood cells and deposition of iron pigments. The deposition of large amounts of iron in hepatic parenchyma causes tissue damage by the enhanced generation of the hydroxyl radical. Excess iron often prediposes to infection. Self-medication with iron-containing tonics by middle-aged persons has been reported to cause toxic symptoms including sideroblastic anaemia, stomach cancer and, sometimes, death. A skin disease called Porphyria cutanea tarda, which is characterized by photosensitivity, blistering, scarring pigmentation and hyperkeratosis, has also been attributed to the overload of iron.

Lead

Toxic symptoms following exposure to lead are manifested at three major sites, namely haemopoietic system leading to anaemia, nervous system leading to palsy, encephalopathy, peripheral neuropathy, learning disability and mental retardation, and kidney leading to renal dysfunction. Furthermore, over exposure to the metal causes gastro-intestinal disturbances, adversely affects reproductive functions and induces mitogenic, mutagenic and teratogenic changes. The absorbed lead accumulates in bones. It may be released later in life due to osteoporosis, which causes delayed toxicity. Deficiency of calcium and iron exacerabates its toxic responses. Age-dependent variations are observed in lead toxicity. Children exhibit higher tendency towards storage of lead compared to adults and are, therefore, prone to lead intoxication. While peripheral neuropathy is more common in adults, encephalopathy is more common in children. The signs and symptoms of its toxicity may be classified as follows:

(a) Gastrointestinal System : The symptoms include anorexia. abdominal pain (colic), vomiting, and constipation or diarrhoea. Presence of a blue line on gums and a metallic taste are diagnostic symptoms of lead intoxication.

(b) Haemopoietic System : Lead interferes with every step of haeme biosynthesis. For example, it reduces the uptake and utilization of iron, enhances biodegradation of haeme, causes haemolysis of mature erythrocytes, lower the concentration of circulating haemoglobin and reduces the mean life-span of circulating erythrocytes in both adults and children. These changes are followed by anaemia associated with reticulocytosis and basophilic stippling of the erythroblastic cells. The biochemical assays, which are considered to be the markers of lead toxicity, include the inhibition of aminolevulinate dehydratase (ALAD) activity in blood and accumulation of zinc protoporpyhrin (ZPP) in erythrocytes. These tests have largely replaced the more complicated determinations of lead in blood and biological samples for diagnostic and screening purposes.

(c) Cardiovascular System : Both animal and human studies show electrocardiographic abnormalities and elevation in blood pressure following lead intoxication. The lower range of blood lead concentrations that may be associated with hypertension have not been precisely described. These effects have been attributed to changes in vascular reactivity, intracellular calcium-ion concentrations, sympathetic tone and alterations in renin-angiotensin system. The exact biochemical mechanisms involed in the toxic manifestations of lead, however, remain to be determined.

(d) Nervous System : The metal elicits electrophysiological, neurochemical and neurobehavioural effects. Of all organ systems, brain reacts most sensitively to lead exposure. The effects may be serious and may lead to paralysis of nerves supplying flexor and extensor muscles of extremities or to

lead encephalopathy characterized by fatal derangement of the central nervous system with convulsions and/or coma. More subtle effects include slowing of the nerve conduction velocity, headache, lassitude, insomnia, fatigue, tremors, nervousness, colic, mania at repeated intervals, cerebral oedema, alterations in mood, psychosis, hallucination, loss of memory and ability to concentrate, I.Q. deficiency, learning disability and mental retardation in children. Though the question of threshold blood-lead levels is still open, concentrations below 40 μg/dl did not provide convincing evidence for congitive effects. Several studies report links between lead exposure and altered electro-physiological and neurotransmitter function involving acetylcholine, norepinephrine, dopamine, 5-hydroxytryptamine, gama amino butyric acid (GABA), calcium, etc.

(e) Kidney : Lead causes nephrotoxicity with impaired reabsorption, porphyrinuria, aminoaciduria, glycosuria, hyperphosphaturia, acidosis, transient decreased synthesis and release of renin followed by a prolonged increase, tubular and vascular lesions and nephrosclerosis.

(f) Miscellaneous Effects : Over exposure to lead is reported to cause endocrine and reproductive abnormalities, spontaneous abortions, decreased fertility, nuclear polyploidy, abnormal mitosis in bone-marrow cells, premature births, foetal abnormalities, immunotoxicity and carcinogenesis, both in man and animals. A characteristic feature of lead toxicity is the occurrence of inclusion bodies in liver cord cells, proximal tubular cells of kidneys, osteoblastic and brain glial cells. These inclusion bodies are believed to maintain low lead concentration in the cytoplasm and thereby serve as a protective mechanism of the body.

Manganese

Feeds containing more than 2000 ppm manganese have been reported to inhibit haemoglobin synthesis by decreasing the intake of iron and its incorporation in the porphyrin ring, causing sideroblastic anaemia, gastro intestinal disturbances, cirrhosis, testicular damage, and degenerative changes in the spinal cord of domestic animals. Elemental manganese is relatively non-toxic but permanganate is strongly oxidizing and a potential toxic agent. A ten gram dose of potassium permanganate is fatal for horses.

Manganese toxicity is a serious threat to health in some industries. The new-born and infants constitute high-risk groups for manganese intoxication because of their low capacity to excrete the metal. The clinical features of its toxicity include broncho and lobar pneumonia not responsive to antibiotic therapy, akinesia, rigidity, increased muscle tone, altered gait, trembling hands, emotional instability, insomnia, hyperirritability, hallucination, mania, violent acts, compulsive phenomena and frank psychosis. These symptoms resemble those of Parkinson's disease. The mental symptoms recover on cessation of its exposure but extrapyramidal abnormalities are usually permanent. The underlying biochemical mechanisms which lead to its toxic manifestations remain to be delineated. Disturbances in dopamine, melanin and possibly serotonin synthesis in basal ganglia have been suggested as the initiating biochemical events. Another disease called sclerosis amyotrophic lateralis, which is reported from Japan, has been linked to the high manganese content in soil and food chain.

Mercury

Symptoms of mercury poisoning include gingivitis, loosening of teeth, metallic taste, stomatitis, gastroenteritis, dysphagia, diarrhoea, cough, chest pain, dyspnoea, paralysis of the muscles of larynx and pharynx, bronchitis, pneumonitis, renal damage (polyuria followed by oligouria and anuria), arrhythmias, cardiovascular collapse, neuropsychiatric manifestations, skin hypersensitivity reactions, immunotoxic, mutagenic, carcinogenic and teratogenic effects. Critical urinary concentrations above which toxic manifestations may be visualized range between 1-2 μg/ml. Oral ingestion of one gram of mercuric salt is lethal.

A dark line of HgS or a dark stippling of gums and ulcers on lips and cheeks may appear.

Mercury intoxication, by both inorganic salts and organic compounds, may lead to neurological damage and grave genetic defects. Alkyl derivatives are more toxic than other forms. The damage caused by such compounds to the nervous system is often irreversible. The Minamata disease, occurring as an epidemic has been related to the organo-mercurial seed dressings and consumption of contaminated fish as observed in Japan and of bread as observed in Iraq. It is manifested by fatigue, loss of memory and concentration, headache, insomnia, drowsiness, numbness in limbs and face, paraesthesia, ataxia, tremours (fine trembling interrupted every few minutes by coarse shaking), dysarthria, constriction of visual field, cortical blindness, hearing defects, acute psychosis, slurred speech, imprecise handwriting with missing words, uncoordinated gait, etc. A typical comatose posture with arched neck, tightly flexed arms, clenched fists and crossed legs have been described. Forced limb extension in such patients is not possible. Methyl mercury-induced behavioural effects are reversible but damage to the foetal nervous system is usually permanent. Principal damage is caused to granular layer in cerebellum and the visual cortex with neuronal loss and glial proliferation. The biochemical site of action is the sulfhydryl group of enzymes.

Cases of sudden death have been reported following i.v. use of mercurial diuretics in congestive heart failure. Death occurs within few minutes of injection. It is believed to be due to ventricular arrhythmias. Cardiovascular collapse is more common in patients with nephrotic syndrome. Atreoventricular dissociation, extrasystoles and ventricular tachycardia precede fibrillation.

The recognized supremacy of non-mercurial preparations has largely replaced mercurials for antiseptic, parasiticidal and fungicidal effects. The alkyl and alkoxy compounds are very toxic, cause local irritation and systemic poisoning when absorbed in the body. Hypersensitivity skin reactions may result from the use of mercurial diuretics in some patients. These may vary from erythema involving the area of contact to severe but generalized eruptions. A mercury-induced syndrome, known as the pink disease, has been reported in children. It is characterized by erythematous fingers, toes, cheeks, nose and buttocks.

Molybdenum

Human molybdenum toxicity is not a major problem. Cases of its occupational exposure in foundry workers have, however, been reported with symptoms of anorexia, pneumoconiasis, elevated blood uric acid levels, fatigue, headache, pain in joints, bone deformities, tremours in hands and disturbances in the central nervous system. The risk is increased in individuals having copper deficiency or fluorosis. Excess molybdenum level has been associated with a higher incidence of gout.

Nickel

Exposure to nickel through dermal, oral or inhalatory routes has been reported to cause allergy, dermatitis, eczema, rhinitis, sinusitis, bronchitis, pneumonitis, asthma, perforation of the nasal septum, inhibition of phagocytosis and the T-cell-mediated immune response, and mutagenic, carcinogenic and teratogenic effects in animals and man. The most common form of human toxicity of nickel is dermatitis. A number of survey reports show 4-13% incidence of sensitivity to this metal.

Platinum

Occupational or therapeutic (as surgical implants and anti cancer cis-platinum compounds) exposure to platinum has been shown to cause allergic dermatitis, urticaria, corrhyza-like syndrome, asthma, dyspnoea, renal damage, and myelosuppression.

Selenium

Experimental toxicity in rats, hamsters and rabbits causes anorexia, enteritis, growth depression, weight-loss, renal and hepatic damage, anaemia, postural defects, paralysis of hind-quarters, reduced fertility and teratogenic effects. Ingestion of selenoamino acids in plants (3-20 ppm Se) has led to acute and chronic toxicity in farm animals. Two distinct diseases related to excessive selenium exposure have been described: (i) blind staggers with symptoms of hypersalivation, garlic-like odour in breath, emesis, diarrhoea, dyspnoea, pulmonary oedema, tachycardia, depression, ataxia, incoordination, paralysis and death, and (ii) alkali disease with predominant hepatotoxic effects, growth depression, low haemoglobin levels, ascites, and atrophy of liver and heart. The internal organs show extensive damage and haemorrhages.

Man is relatively less sensitive to the toxic effects of selenium. Industrial exposure to fumes/dust causes irritation of the mucous membranes of eyes, nose and the upper respiratory tract. This may also be associated with a metallic taste in the mouth, lassitude, fatigue, dizziness, irritability and pyrexia. In some seleniferous areas, such as Enshi region of China, ingestion of food and drinking water with high selenium content result in chronic human selenosis with symptoms of garlic-like odour in breath, bad teeth, icteroid skin, dermatitis, hair loss, diseased nails, gastrointestinal disturbances, inflammation of mucous membranes, numbness, inappetance, headache, heart burn, convulsions, paralysis, etc. Blood and hair selenium concentrations were found to be 30-fold and 100-fold higher respectively in the victims as compared to the control subjects. It is interesting to note that while endemic selenium toxicity has been observd in China, in areas of USA where selenosis was prevalent in animals, no specific signs of the disease were discernible in humans.

Silver

Oral toxicity of silver is low in both animals and man. Rare accidental ingestion of medicinal silver compounds or occupational exposure may cause generalized argyria with a slate-grey appearance of skin, hair and tissues due to deposition of silver. No other symptoms are usually observed.

Thallium

The accidental or homicidal ingestion of thallium-containing pesticides, rodenticides and depilatory agents is the major cause of its exposure to humans. It is manifested by gastroenteritis, mixed neuropathy (motor and sensory), pain, paraesthesia, behavioural disorders, choreform movements, alopecia, vision defects, and damage to the central nervous system which may ultimately prove to be fatal. The effects have been attributed to the formation of an insoluble thallium complex with riboflavin and interference with the sulfur metabolism.

Tin

Inorganic tin compounds are relatively non-toxic. Trialkyl-substituted compounds are the most toxic organotin compounds. Their exposure results in a severe damage to the central nervous system, and it may cause death from paralysis of respiratory muscles. Its widespread poisoning, from the medicinal use of a tin preparation for the treatment of a staphylococcal skin infection has once been recorded in France. The major cause of toxicity was attributed to the presence of substantial amounts of triethyl tin as impurity in the medicine. Cases of food-poisoning due to tin contamination in canned food and fruit juices have been reported with non-specific symptoms, such as nausea, vomiting, diarrhoea, ataxia, etc. Experimental data from laboratory animals indicates the interaction of tin with calcium, copper, iron, selenium, zinc and other nutrients, causing anaemia, disturbances in cellular functions and bone mineralization.

Uranium

It is a highly toxic metal for both animals and man. The toxicity is manifested by dermatitis, bone demineralization, fatal renal damage, and mutagenic, carcinogenic and teratogenic effects.

Vanadium

Because of the low toxicity of ingested vanadium, dietary poisoning does not normally occur. Inhalatory exposure to vanadium fumes causes irritation of the respiratory mucous membranes and conjunctiva, with symptoms of cough, rhinitis, conjunctivitis, chest pain, asthma, bronchospasm, nausea and reduction in the vital capacity of lungs.

Vanadium has been hypothesized as one of the etiological factors of manic-depressive psychosis (MDP). High dietary-intake interferes with the sodium pump in suscpetible individuals, inducing symptoms of MDP. These symptoms are often relieved by the administration of vanadium inhibitors, vitamin C and ethylenediamine tetra acetate (EDTA).

Zinc

Ingestion of large doses causes irritation and corrosive damage to the gastro-intestinal tract, with symptoms of nausea, vomiting, cramps, colic, diarroea, fever and fatal shock. Its toxicity may occur from an intake of acidic fruit juices stored in galvanized, i.e., zinc plated, containers where zinc concentrations may reach upto 2g/litre.

Metal fume fever results from inhalation of more than 5 mg ZnO per m^3 with a particle size of 0.2 to 1 μm. It is attributed to the initiation of an immune reaction and manifested by sore throat, cough, chills, pneumonia, pulmonary oedema, fever, nausea and vomiting. Its chronic human toxicity has not been reported. The therapeutic or prophylactic oral intake upto 350 mg of zinc sulphate does not cause appreciable side effects. Subtle effects on bone development and the immune system with the induction of copper deficiency may be observed.

The principal human diseases due to the deficiency or over exposure of 17 metals/metalloids have been summarized in Table 18.

TABLE 18 METAL-RELATED HUMAN DISEASES

METAL/METALLOID	DISEASES	
	DEFICIENCY	EXCESS
(1)	(2)	(3)
Aluminium	--	Alzheimer's disease Dialysis dementia Dialysis osteodystrophy
Arsenic	--	Black-foot disease
Beryllium	--	Pulmonary granulomatosis
Cadmium	--	Itai-Itai disease
Chromium	Aging (?) Atherosclerosis (?) Diabetes mellitus (?)	Bronchial asthma Contact dermatitis Skin allergies
Cobalt	Pernicious anaemia	Beer drinker's disease Polycythemia
Copper	Menke's kinky hair disease or Trichopoliodystrophy	Grape garden sprayer's disease Indian childhood cirrhosis Schizophrenia (?) Wilson's disease

(1)	(2)	(3)
Gold	--	Aplastic anaemia Mucocutaneous reactions Thrombocytopenia
Iron	Microcytic hypo-chromic anaemia	Haemochromatosis Porphyria cutanea tarda. Sideroblastic anaemia
Lead	--	Lead encephalopathy Lead palsy Learning disability Mental retardation
Manganese	--	Parkinson's disease-like syndrome Sclerosis amoytrophic lateralis
Mercury	--	Minamata disease Pink disease
Molybdenum	--	Gout
Nickel	--	Dermatitis Pneumonia
Selenium	Keshan disease Kaschin-Beck disease	Endemic human selenosis in China
Vanadium	Dental caries(?) Nutritional oedema(?)	Manic depressive illness (?)
Zinc	Acrodermatitis entero-pathica Hypogonadism	Metal fume fever

Note: (?) represents incomplete evidence.

4.5.3 Epidemiological Aspects

Epidemiology is the study of disease as it affects human population rather than individuals. Such an approach avoids complications due to the wide variation in the susceptibility of individuals to toxicants because of variability in age, sex, race, diet, drugs, and inherent differences in absorption and adoptation. These studies have fruitfully linked low levels of many toxic metals in the environment or in the tissues of subjects to particular disease states. Some interesting examples are given below.

Arsenic

Epidemiological studies have demonstrated a certain association between environmental, occupational, and medicinal exposure of man to inorganic arsenic and cancer of skin and lungs. The time-lag between the first case reports of arsenic-related cancer and its epidemiological confirmation was about 70 years. The use of sodium arsenite as herbicide is forbidden in many countries.

Beryllium

Epidemiological studies are difficult because only a small fraction of the exposed population develops symptoms of beryllium disease. The time-lag between the first indication of carcinogenic effects of beryllium and its confirmation in animal studies was only seven years. The effects reported in human subjects were less pronounced but were discernible even 30 years after the exposure.

Cadmium

Cadmium levels in human blood and tissues are highest among the Japanese. The Itai-Itai catastrophe at the Jintsu river has been connected with cadmium-rich effluents from an abandoned lead-zinc mine. The disease has been attributed to the consumption of contaminated drinking water and of rice crop irrigated with such water. Though visible damage to the rice crop was observed earlier, it took 17 years from the first medical studies to the final fixing of legal responsibility on the mining company that discarded the cadmium effluents into the river. The disease was further linked to a concurrent calcium deficiency as it affected older women and caused bone decalcification and frequent fractures. Correlations with animal data proved of immense help in establishing the etiological association.

Chromium

Dermatitis and eczemas caused by contact with cement have been linked to allergic reactions against Cr (VI) impurities in cement. In a Scandinavian study on a population group of over 5500 patients, 3% of the women and 12% of the men with skin allergy, showed positive reaction tests with potassium dichromate. Epidemiological studies have also demonstrated an association between occupational exposure to this metal and lung cancer. The time-lag between the first case reports and its epidemiological confirmation was about 20 years.

Lead

Industrialization and increased transport activities have led to lead pollution of the natural environment on a global scale. The metal has been shown to impair neurological, cardiovascular and reproductive functions and cause carcinogenic effects. According to a UNEP/WHO report, vast variations are found in the blood-lead levels among the inhabitants of big cities of the world, e.g., Tokyo (60 μg/1) and Mexico City (225 μg/1). This is alarming since at blood levels > 80 μg/d1, there is high risk of adverse cardiological effects. The main source of lead pollution is air from lead smelters, industrial centres and traffic inputs. About 1 μg/m^3 of lead in air is reported to account for 2 μg/d1 of the metal in blood. A curvilinear relationship between dietary and blood-lead concentrations is also noted.

Children are particularly prone to lead intoxication. A significant correlation exists between blood-lead concentrations of mother and children. Epidemiological studies conducted in Britain, Germany and USA have shown a causal relationship between exposure to lead and deficits in cognitive functions. Children, as a risk-group for central nervous system effects, have received particular attention in such investigations. The results indicated neuropsychological deficits and lower I.Q by 5-6 points at blood levels >40 μg/dl. No evidence was, however, discernible to link moderate lead exposures and behavioural patterns in general. Though, an interplay of several genetic and environmental factors may be involved, a population shift of 5-6 I.Q. points cannot be ignored.

Several studies report increased mortality among lead smelters and battery plant workers from gastro-intestinal, respiratory and renal cancers. Exposure to lead for more than 20 years may lead to increased incidence of cancer. Though some studies have been criticized on methodological grounds, the data so generated is certainly a cause for concern and calls for adequate steps for predictive and preventive measures.

Mercury

The Minamata disease is a classical example of poisoning with methyl mercury on an epidemic scale. The first case was reported by Dr. Hosokawa in 1956 near Minamata bay in southern Japan. It was shown to be caused by eating fish contaminated with CH_3HgCl discharged by a chemical plant into the bay. The mercury load was so great that it killed fish and fish-eating birds, which dropped into the sea with intense symptoms of violent whirl, dashing with a zig-zig course and collapse. Though links between water pollution, consumption of contaminated fish and human disease were established in 1959, the effluent dumping could be halted nine years later in 1968.

Consumption of bread contaminated with mercury-containing fungicides led to intoxication in over 60,000 people, more than 6000 hospital admissions and 200 deaths in Iraq in 1971-72. The symptoms resembled those of the Minamata disease. Similar epidemics were reported from the Nigata region of Japan, Canada, USA and some other countries. Populations exposed to lower levels of methyl mercury in Canada and New Zealand (through fish consumption) exhibited abnormal muscle tone, deep tendon reflexes and poor performance by children in intelligence tests. The Environmental Protection Agency prohibited registration of methyl mercury seed dressings fungicides in 1972.

Nutritional contamination with organic mercury is highly neurotoxic, specially for young children and pre-natal development of foetus. Children born to mildly diseased mothers (with or without symptoms) suffered from cerebral palsy and idiocy. Women with more serious diseases showed higher incidence of abortions and still births. Epidemiological studies have also shown links between mercury exposure and leukemias. Methyl mercury aggravates the toxicity of some pesticides. It is surprising to note that some healthy children from Greenland had more than 170 µg/1 of mercury, which is almost equal to the blood levels of victims with congenital Minamata disease. Selenium elicits a protective role against mercury intoxication. More research is needed to probe these aspects.

Nickel

Epidemiological studies showed high incidence of lung, larynx, and sinus cancers among workers exposed to nickel in refinery, smelting and electrolysis works.

Selenium

Studies in population groups from 27 countries showed an inverse relationship between environmental selenium concentrations/dietary intake and incidence/mortality due to cancer at 17 major body sites, including colon, rectum, lung, breast, ovary and blood. Geographical association between selenium status and cancer mortality was recently confirmed through studies in China and USA. Cases of chronic endemic selenotoxicosis have also been reported from China and Latin America. These were attributed to consumption of food with high selenium content. Selenosis occurs from industrial exposure.

Zinc

Epidemiological studies have established a definite relationship between zinc deficiency and male hypogonadism. Some investigators reported a positive correlation between zinc in semen and fertility status. Studies showing no significant difference in semen-zinc content from fertile and infertile men are available. At present, many questions remain unanswered regarding the mechanisms involved and the role of zinc in sperm motility, maturition, its journey through the female genital-tract, its viability, as well as its therapeutic efficacy in clinical situations.

The above examples are obviously illustrative and not exhaustive. An epidemiological approach may help us to identify: (i) hazards associated with other heavy metals not previously considered as a problem, (ii) new aspects of known metal-related diseases, (iii) particular source(s) of exposure, (iv) critical organ/tissue involved, (v) high-risk group, (vi) critical concentrations/tolerance limits, (vii) useful bioindicators, and (iii) ways and means for ecological prevention and effective control of ill-effects of metal exposure.

The studies involving exposure effect or human response obviously need meticulous planning and execution by efficient personnel with expertise, keen observation and dedication. The studies should include population groups with high risk, women in reproductive years, young children, ambulent workers and patients with manifest intoxication. Quantitative assessment of the exposure limit is an important aspect. In addition to the customary recording of morbidity and mortality data and observations of gross symptoms, the emphasis now is on the detection of subtle pre-clinical effects by behavioural investigations and on the use of biochemical indices of alterations in body functions. The diagnosis of low-level chronic metal toxicity by simple practical methods has largely been frustrating and needs constant attention for obtaining meaningful results.

4.5.4 Therapeutic Aspects

Therapeutic aspects will be discussed under two heads:
1. Treatment of heavy metal poisoning
2. Medicinal uses of heavy metals

1. Treatment of Heavy Metal Poisoning

The general measures of treatment and control of heavy-metal poisoning include:

(i) Cessation of exposure.

(ii) Decontamination measures, e.g., washing of skin, cathartics (sodium or magnesium sulphate), induction of emesis (ipecac), prevention of absorption (milk, egg-white, tannins), adsorption (activated charcoal), use of diuretics after prior hydration, etc.

(iii) Elimination enhancement by haemodialysis, peritoneal dialysis, exchange transfusion, and haemoperfusion.

(iv) Chelation therapy using desferrioxamine (DFO), disodium calcium ethylenediamine tetra acetic acid ($CaNa_2$ EDTA), British anti-lewisite (BAL), d-penicillamine, etc.

(v) Supportive care includes hospitalization of the patient, avoiding chills, keeping the body warm, immediate removal of obstructions to breathing, cleaning of respiratory airways, institution of endotracheal intubation, artificial respiration and/or oxygen, if necessary, maintenance of fluid and electrolyte balance, careful monitoring for vital signs, renal and hepatic function, ECG, pulse, blood pressure, hypovolaemia, shock, dyspnoea, and blood chemistry, adequate nutritional status (high-protein diet enriched with vitamins, calcium, iron, etc.), and a careful watch on relapse.

(vi) Use of specific antidotes.

(vii) Symptomatic treatment to control concurrent infection, inflammation, allergy, shock, bronchospasms, convulsions, coma and burns by appropriate drugs including antibiotics, corticostriods, antihistaminics, analgesics, sedatives, analeptic, emollients, etc.

Specific measures for poisoning with individual metals/metalloids are presented in Table 19.

TABLE 19 TREATMENT OF HEAVY METAL POISONING

METAL/METALLOID	ANTIDOTE/THERAPEUTIC MEASURES
Aluminium	(i) Chelation therapy with desferrioxamine (DFO)
	(ii) Reverse osmosis and deionisation of water for dialysis.
	(iii) Aluminium-containing medication to be avoided in patients undergoing dialysis.
Antimony	(i) British anti-lewisite (BAL).
	(ii) Dialysis needed in case of stilbine gas exposure.
	(iii) Monitor for arrhythmias.
Arsenic	(i) Urine arsenic levels >200 μg/litre indicate chelation with BAL or penicillamine; the latter is preferred.
	(ii) Unithiol, dicaptol, dithioglycerol, dimercapto succinic acid, sodium thiosulphate and ascorbic acid are useful.
	(iii) Haemodialysis if renal failure coexists.
	(iv) In case of arsine gas poisoning, if free haemoglobin levels exceed 1.5g/dl, exchange transfusion may be given.
	(v) Monitor for underlying diseases and complications such as anaemia, ECG changes, hepatic enzymes, pulmonary function, hyperkalemia, etc.
Beryllium	(i) Chelation therapy is usually ineffective in removing Be^{2+} from tissues.
	(ii) Aurin tricarboxylic acid.
	(iii) Symptomatic therapy using antibiotics, corticosteroids, bronchospasm-arresting drugs, cardiac glycosides, diuretics, oxygen, etc.
Cadmium	(i) Though chelating agents, such as calcium ethylenediamine tetra acetate

METAL/METALLOID		ANTIDOTE/THERAPEUTIC MEASURES
		(Ca EDTA) have been used, they may harmfully redistribute Cd^{2+} to kidneys. Furthermore, a delayed treatment is not much effective.
	(ii)	BAL increases nephrotoxicity and, therefore, has not been recommended for use. Chronic exposure requires no chelation.
	(iii)	Charcoal is not useful.
	(iv)	Unithiol
	(v)	High-protien diet with adequate Ca, Zn, phosphate and vitamin D content reduces toxicity.
Chromium	(i)	Removal of contaminated clothing.
	(ii)	Washing of skin with water and application of dressings soaked with EDTA emulsion.
	(iii)	Efficacy of BAL, ascorbic acid, folic acid, haemodialysis and exchange transfusion has not been established.
	(iv)	Unithiol
	(v)	Monitor renal and hepatic function.
Cobalt	(i)	Disodium salt of EDTA or trisodium calcium salt of diethylene triamine penta acetic acid (DTPA).
	(ii)	Antihistaminics and other symptomatic measures.
Copper	(i)	Disodium calcium EDTA i.v. or BAL i.m. are the agents of choice in severe poisoning.
	(ii)	D-penicillamine may be administered p.o. but it may induce allergic reactions in some cases.
	(iii)	Exchange transfusion with albumin-enriched peritoneal dialysis.
	(iv)	Molybdenum and zinc salts may help reduce toxicity.

METAL/METALLOID		ANTIDOTE/THERAPEUTIC MEASURES
Gold	(i)	Chelation with BAL.
	(ii)	Steriods for pneumonitis and supportive care.
Iron	(i)	DFO is a specific chelator. Iron-desferrioxamine complex is highly stable at physiological pH. This helps in keeping the free iron in the extra-cellular space and reduces the disruption of cell membranes and enzyme systems. DFO may be administered at a dose 15 mg/kg/h i.v. or 40-90 mg/kg giving a deep i.m. injection.
	(ii)	Exchange transfusion.
	(iii)	Haemodialysis removes iron-DFO complex but not iron itself. Its use has been recommended in patients with impaired renal function.
	(iv)	Gastrotomy reduces toxicity.
	(v)	Watch for signs of hypovolemia and monitoring for prothrombin time, transaminase and creatinine levels has been recommended.
Lead	(i)	Chelation with disodium calcium EDTA, BAL-Ca EDTA combination or d-penicillamine. Indications for chelation include acute encephalopathy, classic plumbism and some asymptomatic cases with high lead levels. Disodium calcium EDTA should not be administered orally as it enhances lead absorption through this route. D-pencillamine can mobilize bone lead when given orally. Combined therapy with Ca, Na_2 EDTA and BAl is quite effective in preventing rapid mobilization of lead into blood. Procain or lidocaine may be added during i.m. injection to reduce pain.
	(ii)	Unithiol

METAL/METALLOID	ANTIDOTE/THERAPEUTIC MEASURES
	(iii) Monitor blood-lead levels, watch for deteriorating renal function and maintain generous urine flow (350-500 ml/m^2/d).
	(iv) Diet with high calcium, adequate iron and low phosphate levels and dust control measures are helpful.
	(v) Symptomatic treatment for cerebral oedema and encephalopathy.
Manganese	(i) No universally effective antidote is available.
	(ii) Calcium EDTA administration in conjunction with supportive treatment may be given.
	(iii) Administration of L-3, 4-dihydroxyphenyl alanine (L-DOPA) in early stages of the disease provides symptomatic relief.
Mercury	(i) BAL is recommended in cases of severe inorganic mercury poisoning. Its use and efficacy in organic and elemental mercury poisoning is questionable. Simple administration of BAL should be avoided as it facilitates the distribution of mercury throughout the body. BAL-Hg complex causes renal damage.
	(ii) Haemodialysis, haemoperfusion and peritoneal dialysis remove negligible amounts of the metal since a little mercury is available unbound in the plasma. Haemodialysis with infusion of cysteine or N-acetyl penicillamine is useful. D-pencilliamine reverses SH-binding in the blood and chelates the metal. Its use is contraindicated in penecillin-sensitive patients.
	(iii) Unithiol
	(iv) Administration of polysaccharides, lipotropic agents, vitamin B_{12} and folic acid is beneficial.

METAL/METALLOID	ANTIDOTE/THERAPEUTIC MEASURES
Molybednum	(i) Unithiol, methionine, and sodium sulphate are useful. (ii) Copper-rich diet reduces molybednum toxicity. (iii) Symptomatic treatment using antibiotics, cardiovascular drugs, vitamins, etc. may be given.
Nickel	(i) Chelation with BAL. (ii) Unithiol (iii) Lanolin castor ointment with EDTA protects the skin. (iv) Symptomatic treatment should be given for inflammatory states of the respiratory mucosa (anti-inflammatory agents/antibiotics) and pulmonary oedema (blood-letting, diuretics).
Platinum	(i) Control of dust exposures and thorough washing of the affected area reduces the chances of industrial allergy. (ii) Symptomatic treatment includes the use of osmotic diuretics with vigorous prehydration, adrenaline injections, oxygen administration, nervine tonics, vitamins, etc. Cisplantin-induced ototoxicity (tinnitus and hearing loss) is generally irreversible.
Selenium	(i) Treatment of patients exposed to compounds containing selenious acid (e.g., gun-bluing solution) has not been well studied and no specific antidotes are available. (ii) Elimination of the selenium source. (iii) Desensitizing drugs. (iv) Application of 10% sodium thiosulphate solution on burns caused by selenium dioxide. (v) BAL and $CaNa_2EDTA$ should be avoided as these may enhance selenium toxicity.

METAL/METALLOID		ANTIDOTE/THERAPEUTIC MEASURES
	(vi)	Symptomatic measures such as cardiopulmonary monitoring under intensive care setting, i.v., fluids, supplementary oxygen, etc. are helpful.
Silver	(i)	Dimercaprol is ineffective. There is no specific antidote.
	(ii)	Symptomatic and supportive measures are recommended.
Thallium	(i)	Dimercaprol and EDTA are ineffective.
	(ii)	Activated charcoal binds the thallium ion. Gastric lavage with water containing activated charcoal and 0.3% sodium thiosulphate solution is useful.
	(iii)	Potassium iodide diuresis increases thallium excretion.
	(iv)	Prussian blue, dithizone, haemoperfusion, diathermy, oxygen, etc. may be used.
Tin	(i)	BAL does not chelate organic tin effectively. D-penicillamine does not enhance urinary elimination of tin.
	(ii)	Tin poisoning is managed largely by supportive and symptomatic measures.
Uranium	(i)	Chelation with calcium EDTA.
	(ii)	Use of bicarbonates, citrates and other supportive measures.
Vanadium	(i)	Chelation with DTPA
	(ii)	No specific antidote is available. Use of vitamins B and C is beneficial.
Zinc	(i)	Disodium calcium EDTA is an agent of choice. It normalizes zinc levels.
	(ii)	D-penicillamine has been shown to increase mortality in animals. Human data is not available.
	(iii)	Use of alkaline inhalations, unithiol, glucose, vitamin C, tea/coffee, cardiac drugs, and antibiotics is indicated according to the symptomatic demand.

2. Medicinal Uses

So far, the deleterious effects of heavy metals have been documented. However, their therapeutic potential in disease states has also been exploited. Some such examples of the medicinal use of metals/metalloids are described in short below :

Arsenic

Recently, clinical trials were conducted at the Maulana Azad Medical College and Associated Hospitals, New Delhi, on the usage of Kushta Hartal (a calcined arsenic preparation used in Unani medicine and claimed to be an effective and safe drug for the treatment of diabetes mellitus). Comparative evaluation was performed on 206 uncomplicated cases of maturity-onset diabetes in two groups of matched age, sex, and intensity of diabetes status using this drug (103 patients) and chlorpropamide (103 patients). In the Kushta Hartal group, 90/103 patients achieved adequate control followng treatment with a daily dose of 500 mg and 13 were controlled with 250 mg. In the chlorpropamide group, 80 patients were controlled with doses ranging between 230-375 mg per day while 23 required only 250 mg per day. Symptomatic improvement along with blood-sugar control achieved in both the groups, was not found to be significantly different. The investigators also reported an excellent tolerance to Kushta Hartal whereas normoglycemia and freedom from diabetic symptoms were maintained during a follow-up period of two years.

Arsenic preparations have also been used for the treatment of leukemias, amoebiasis, trypanosomiasis, syphilis and skin diseases. The metal, however, has limited application and its use has largely been abandoned following the availability of better and less toxic drugs.

Chromium

An impairment of intra-venous glucose tolerance has been observed in chromium deficiency. It can be prevented through the dietary supplementation of this metal. In a study conducted on 12 malnourished children (6 each from Jordan and Nigeria) following chromium therapy with 250 µg given as its chloride, the i.v. glucose tolerance tests were found restored to normal levels. Improvement in the glucose tolerance of maturity-onset diabetes among all age-groups has been reported in 50-60% of the patients receiving a prolonged chromium therapy. Hyperglycaemia in patients kept on prolonged parenteral nutrition has been controlled by oral administration of chromium in microgram quantities and their insulin requirements have also been reduced.

High-chromium yeast has been reported to be useful in the treatment of acne. This is not surprising because both insulin and tolbutamide elicit beneficial effects in acne. Probably, the therapeutic efficacy of chromium in acne is mediated through its effect as a glucose tolerance factor.

According to an Indian study, patients sufferings from nervous system weakness (due to excess physical and/or mental strain) get relief with regular intake of 5 drops of 0.05 M chromium sulphate twice a day after meals. The investigators further claimed that a nervous-system disability originating from gastric hyper-acidity and gastric distress following gas retention is considerably relieved after Cr^{3+} therapy. These studies suggested its possible therapeutic application. However, further detailed investigations are needed before it may be taken as valid for human use.

Cobalt

The best example of the potential therapeutic use of cobalt is as cyanocobalamine (vitamin B_{12}), which is given for the treatment of pernicious anaemia.

Copper

Copper compounds are used as fungicides, insecticides, and pesticides while its metallic devices are used as intra-uterine contraceptives. Copper complexes with other anti-inflammatory drugs (e.g. niflumic acid and phenylbutazone) have been shown to elicit better action than when they are used alone.

The studies conducted at a Delhi-based skin institute revealed low serum-copper values in vitiligo patients. The therapeutic use of copper chloride, 1-3 mg/day in capsules or 200 μg i.v. twice a week, proved effective in patients suffering from this disease. Though the precise etiology of vitiligo is not understood, the disease has been attributed to a disturbance in the synthesis of melanin that involves tyrosinase, a copper-containing enzyme. It has been hypothesized that supplementation with exogenous copper may help in correcting the metabolic deficiency of the metal. Clinical evidence on therapeutic efficacy is so far inadequate.

Copper bracelets are traditionally claimed to be of therapeutic value in arthritic patients. It is believed that the metal gets dissolved in the sweat and enters the circulation. These claims have never been verified scientifically.

Gold

Gold was used in the treatment of rheumatoid arthritis as early as 1929. The treatment requires careful supervision. The most serious effect of gold therapy is the suppression of production of cells such as platelets, and red and white blood cells in the bone marrow. Its complete suppression can lead to aplastic anaemia, which may sometimes be fatal. Besides, therapy with this metal is not effective in all cases of rheumatoid arthritis. The relapse rates are usually very high. In general, out of every 100 patients of rheumatoid arthritis under treatment with gold compounds, twenty will get no benefit, 25-40 will develop toxic reactions (half of them sufficiently serious to require discontinuation of therapy), and the remaining will improve. Curcumine-gold complex, developed by some Indian scientists from gold and an anti-inflammatory principle of turmeric (*Curcuma longa*), has recently been reported as an effective drug for this disease.

Earlier gold compounds were used for the treatment of many other diseases such as asthma, leprosy, syphilis, tuberculosis, etc., but without any remarkable success. Such uses have now been abandoned and are of historical interest only.

Iron

The use of iron preparations in iron-deficiency anaemias is well known. The usual therapeutic dose is 60 mg. of ferrous iron given three times daily. The daily uptake of 0.2 g of iron increases the haemoglobin content, which may be observed day-3 onwards. The treatment may be continued until the haemoglobin is normalized. Iron can also be given in the form of imferon intravenously. Oral therapy is preferable to parenteral administration. The latter should be avoided as far as possible. Mere iron supplementation may not be enough in case of anaemias. Other factors such as causes of its deficiency, proper utilization, total iron-binding capacity, etc., should also be looked into.

A double-blind random allocation study was conducted in 44 healthy non-aneaemic (haemoglobin >12 g/dl) women complaining of chronic fatigue without headache, dizziness or nervousness. It revealed significant improvement in the group receiving 300 mg of ferrous sulphate thrice daily as compared to the group receiving look-alike placebo tablets administered in a similar manner. Iron is not a general tonic. Its use in ailments other than iron deficiency states does no good to the patients and may even be harmful. The bacteriostatic property of iron-binding proteins (transferrin and lactoferrin) is lost when these proteins are saturated with iron. Its parenteral use can predispose the patients to infection.

Manganese

A South African diabetic, who was poorly controlled by insulin, treated himself with alfa-alfa tea extract. The latter is known to contain a high content of manganese. It was further observed that oral administration of 5 mg of manganese to this patient was effective in restoring normoglycemia within 3-4 h. Further investigations on this aspect may yield some fruitful results.

Mercury

Inorganic as well as organic mercury compounds are used as preservatives, antiseptics, parasiticides, fungicides, and powerful diuretics. The earlier use of mercury preparations as cathartics and in syphilis have been discontinued.

Platinum

Cisplantin and other platinum complexes are reported to elicit antibacterial, antiviral and antitumour activity both in animal models and in man.

Selenium

A four-year study involving more than 46,000 children in the Sichun province of China revealed a significant protective-effect of selenium supplementation (0.5-1.0 mg once a week) against the onset of Keshan's disease in respect to both the incidence and mortality rates.

A recent study described a dramatic improvement in a patient of congestive heart failure following selenium therapy. This 61-year-old patient was on total parenteral nutrition (TPN) for 8 months because of a cancerous obstruction in his bowels. The addition of selenious acid (40 μg/day) in the TPN solution, exhibited marked clinical and radiological improvement in the patient within a week. His serum-selenium levels rose from 0.11 to 0.14 mg/1 whereby digoxin doses could be reduced to 0.25 mg p.o. with adequate serum concentrations. Although these observations do not prove that selenium improves the cardiac function, the report definitely suggests further investigations on its utility in long-term TPN patients and also in enhancing the efficacy of digoxin.

Observation of low serum levels in arthritis patients with no corresponding decrease in its intake prompted a clinical study. Thirty six patients with this disease were given a preparation containing 100/μg organic selenium, 90 mg ascorbic acid, 50 mg d-alpha tocopherol and 1500 i.u. retinol per tablet. Both subjective and objective improvements have been observed in 45.5% patients following a three-month course of two tablets per day during the first and one tablets per day during the next two months.

Selenium is useful in the cure of skin diseases. Selenium sulphide is applied in seborrhoic dermatitis and incorporated as an anti-dandruff agent in shampoos. The anti-dandruff efficacy of two shampoos containing 2% micronazole nitrate (15 patients) and 2.5% selenium di-sulphide (8 patients) was compared and found to be similar. Selenium sulphide (2.5% lotion) applied daily for 10 min for seven consecutive days was shown to be an effective therapeutic agent in tinea versicolor. In a double-blind clinical study, selenium treatment was found to be significantly superior to the placebo group receiving vehicle alone. Mild transitory contact dermatitis of primary irritant type, noted in both groups, was attributed to the detergent base. Another study described selenium sulphate lotion as an effective adjunct to griseofulvin in the treatment of tinea capitis. The determination of blood glutathione peroxidase (G-Px) levels in 61 healthy subjects and 506 patients with various skin diseases revealed depressed levels in psoriasis, eczema, alopic dermatitis, vasculitis, mycosis fungoides, dermatitis herpetiformis, pamphigoid, acne conglobata, scleroderma and lupus erythematosis. Fifty six patients with low G-Px levels were treated with tablets containing 0.2 mg selenium as sodium selenite and 10 mg tocopherol succinate. Encouraging clinical results were obtained with concurrent gradual increase in G-Px values within 6-8 weeks of treatment. In another open trial, 29 patients having moderate to severe acne, were given the same treatment twice daily for 6-12 weeks. Promising results were obtained specially in patients with pustular acne and low G-Px activity.

Silver

Silver sulfadiazine and some other complexes of this metal have been attributed to have antibacterial, antiviral and anti-tumour effects. Silver nitrate, allantoinate and sulfadiazine compounds are applied locally for antiseptic and caustic effects. A detailed study on 70 patients demonstrated that 0.5% silver nitrate dressings, changed 12-hourly, is an effective treatment for major thermal burns, wounds and sepsis. The study covered patients of varied age groups (<10 to >50 years) with second and third degree burns affecting 15-92% (mean: 32%) surface area. The treatment showed reduction in the number and variety of organisms on culture, disappearance of purulent exudate and odour from wounds, good healing with fat, firm and bright red granulations, absence of pain, argyria, sepsis and scarring, and low mortality. Traditional use of silver for the treatment of epileptics and lunatics had no rational basis and, therefore, has, been abandoned.

Tin

Tin compounds are used in tape-worm infestations.

Vanadium

The use of vanadium in anaemias, diabetes, tuberculosis and as an antiseptic and spirochaetocidal agent has been abandoned following availability of effective and safe drugs. The potential therapeutic use of vanadates as insulinomimetic and hypocholesterolemic agents is under investigation.

Zinc

Zinc has been put to therapeutic uses in a wide variety of human ailments :

(i) Acrodermatitis enteropathica : It is a well-known inherited zinc deficiency disorder and the fact that it responds to oral zinc supplementation is well documented. In a six-month-old

Saudi boy, cerebral atrophy was completely reversed after oral treatment with zinc sulphate. On the other hand a four-month-old Japanese baby, who was given lactose-free milk (with low zinc content : 0.5 mg/100 g) for about three months to treat diarrhoea, developed this disease. The lesions disappeared on switching to a general milk formula (containing 1.2 mg zinc/100 g) without additional zinc supplementation. Percutaneous absorption of zinc has also been reported in acrodermatitis enteropathica. An eight-month-old girl suffering from this disease, was treated with 25% zinc oxide paste applied locally in the diaper area. A rapid improvement in the symptoms was observed a week after the treatment. Zinc supplementation has also been shown to improve the general conditions including height for age scores in children. If untreated, the disease is often fatal. In cases with severe hypozincemia, parenteral therapy may be necessary.

(ii) Infertility and Male Hypogonadism : In an excellent study, experimental zinc deficiency was induced in male volunteers by restricting the dietary intake of zinc (2.7-5.0 mg/day for 20-40 weeks). This led to the development of oligozoospermia (total sperm count: 40 million per ejaculate) in 4/5 subjects. This adverse effect on testicular function was shown to be reversed by zinc supplementation in physiological amounts. Another study probed the effect of zinc therapy on plasma testosterone (T), dihydrotestosterone (DHT) and sperm count in 37 patients with idiopathic infertility of more than five years duration. In one group ($T < 4.8$ ng/ml; 22 patients), T, DHT and sperm-count rose significantly following administration of zinc. The wives of nine of these subjects became pregnant, six within three months and three within two months of a second trial. In the second group ($T > 4.8$ ng/ml, 15 patients), T and sperm-count were unaffected by zinc therapy while DHT rose significantly. No conceptions were observed in this group. A double-blind study on six zinc sulphate-treated and seven placebo-treated patients, however, did not find zinc to be an effective treatment for hypogonadism and impotence associated with chronic stable hepatic cirrhosis. Twenty stable uraemic patients, undergoing haemodialysis thrice a week, completed a double-blind

trial using either 50 mg of elemental zinc given as zinc acetate (10 patients) or placebo (10 patients) orally for six months. A significant increase in mean plasma zinc, serum T levels and sperm count, coupled with a significant fall in serum LH and FSH concentrations, improved libido and higher frequency of intercourse was observed in the treated group. The results indicate that zinc deficiency is a reversible cause of gonadal dysfunction in patients undergoing regular haemodialysis.

(iii) Atherosclerosis : Some clinical studies indicate beneficial effects of zinc therapy in atherosclerosis. Zinc sulphate, given orally to patients with advanced vascular disease, elicited marked improvement in 12/13 cases. In another study, 16 patients (14 with inoperable symptomatic atherosclerosis and two who had a disabling vasospastic disease) were given zinc sulphate for 3-11 months. Six of these patients were excluded from the study because of observed changes in their diet or habits. Clinical improvement, observed in six of the remaining 10 patients, was attributed to zinc therapy. No improvement was observed in four patients and two showed a rapid progression of the disease.

(iv) Arthritis : In a preliminary trial, oral zinc supplementation yielded encouraging results in 24 patients with chronic refractory rheumatoid arthritis. Zinc sulphate (220 mg three times daily) or look-alike placebo capsules were added to the pre-existing therapy for 12 weeks. This double-blind trial was followed by an open 12-week period when all subjects took zinc. During the double-blind phase, zinc-treated patients felt better than controls with regard to joint swelling, morning stiffness, walking time and patient's own assessment of the disease status. The indices and joint tenderness also improved following zinc therapy in both groups during the second 12-week period. No beneficial effects could, however, be demonstrated on grip strength, haemoglobin content, erythrocyte sedimentation rate or titre of rheumatoid factor. The frequency of headache, rash, anorexia, abdominal pain and diarrhoea was higher in the placebo group. Mild nausea, observed in some cases, could be circumvented by taking

zinc post meals. Beneficial effects of zinc therapy in psoriatic arthritis have also been reported.

(v) Hepatic disorders : Clinical improvement with oral zinc therapy has been observed in some studies on patients suffering from Indian childhood cirrhosis and alcoholic cirrhosis.

(vi) Common cold : In a study on 146 volunteers afflicted with common cold, the patients were given 180 mg of zinc gluconate (containing 23 mg of elemental zinc) or fake look-alikes. The pill was recommended to be sucked for at least 10 minutes so that zinc could give a lengthy bath to the throat tissues where the rhinoviruses multipy. Recovery was achieved in an average period of 3.9 days in the treated-group as against 10.8 days in the placebo-group, indicating a significant therapeutic effect of zinc treatment in common cold.

(vii) Skin diseases : Zinc therapy has been shown to be beneficial in several skin ailments including acne, alopecia, burns, dermatitis, eczema, frunculosis, herpes infection, keloids, leprosy, psoriasis, scars, ulcers and wounds. The metal elicits its curative effects by both oral treatment and local application using gels, ointments, lotions, adhesive tape and dusting powder.

(viii) Complications of Pregnancy : The administration of zinc sulphate (Solvezine 200 mg twice daily) during the third week of pregnancy, was reported to reduce the incidence of foetal and maternal complications without any apparent side effects. In another study, the same treatment was given to 64 women in the fourteenth week of gestation in an open randomized trial. The therapy reduced the number of complications; 63% women delivered normally and got normal babies in the treated group as against 48% in the low-zinc control group. Some Russian workers have reported, successful results with a combination therapy (Zn 5 mg, Mn 4.5 mg, Mg 6.2 g daily) in anaemic pregnant women. The patients required fewer surgical interventions at delivery and showed

reduced frequency of delayed labour besides, better maternal blood values and reduced incidence of low birth weight babies.

The indiscriminate use of zinc during pregnancy must be avoided. In India, a therapeutic trial using zinc sulphate (100 mg per day) during the third trimester of pregnancy was discontinued when four consequent births were premature and one infant was still born. The recommended daily allowances for zinc are 20 mg during pregnancy and 25 mg during lactation.

Miscellaneous

Metals as such or in the form of alloys are used in dental amalgams and surgical implants. These include aluminium, chromium, copper, gold, iron, manganese, mercury, molybdenum, nickel, platinum, silver, tin, titanium, tungsten, zinc, etc.

Complexes of blemomycin (BLM) with cobalt, copper, iron and zinc have been shown to be of potential value as anti-infective and anti-cancer drugs. A study of these complexes in microbial cultures and for anti-tumour activity revealed the following order of their efficacy :

Cu (II) BLM>BLM>Zn (II) BLM> Fe (III) BLM > Co (II) BLM

4.6 TOXICITY

4.6.1 Factors Affecting Toxicity

Factors affecting toxicity may be broadly classified into two groups : metal-related, and host-related. Metal-related factors include abundance, chemical form (elemental, inorganic or organic), speciation, oxidation state, ionizability, particle size, magnitude/duration of exposure, and irritant, corrosive, cytotoxic, mutagenic and carcinogenic properties of the metallic compound. Host-related factors include susceptibility, route of exposure (oral, inhalatory, dermal, parenteral, implants, etc.), bioavailability, absorption, binding, metabolism, excretion, ability to penetrate blood-brain and placental barriers, interactions with dietary nutrients, other elements, alcohol and drugs, critical concentration, nutritional and immune status, target organ/tissue, and critical effect/host response. Details of these factors with specific examples have been discussed in relevant sections of this book. However, some of the important effects which have not been described in detail previously are discussed in the forthcoming sections.

4.6.2 Carcinogenic Effects

A close relationship exists between mutational changes of the genetic material caused by metals and malignant transformation. The ability of a metal to enter the cell, the capacity of its ion to attain significant intra-cellular concentrations to react with DNA, mutagenicity and cytotoxicity are important molecular features of metal carcinogenesis. The following metals/metalloids have been shown to induce chromosomal damage in human cell cultures : As, Cd, Cr, Hg, Ni, Pb, Pt, Se, and Zn. The damage has also been noted in *in vivo* studies for most of these metals except for platinum and selenium. Many of the alleged carcinogenic metals may not be true carcinogens but may act non-specifically as irritants. Many of them, when implanted into muscles, can induce sarcomas possibly as a result of causation of local hypoxia or anoxia. Normal myocardial cells, maintained in tissue cultures under hypoxic conditions, get transformed into malignant cells. Unlike in the case of organic chemicals, the initiation of neoplasia by metals does not necessarily require metabolic activation beyond redox reactions or particle dissolution. Very reactive and highly cytotoxic metal ions such as Hg^{2+} may cause cell-death before any potential carcinogenic response may become apparent. A brief discussion on the role of individual metals/metalloids in carcinogenesis is given as follows :

Aluminium

Exposure to aluminium dextran and aluminium oxide dusts (particle size : 1-5 micron) has been shown to be carcinogenic in mice. However, epidemiological studies and mutagenic tests in bacteria did not indicate such effects with aluminium compounds.

Antimony

A high incidence of urinary bladder tumors, seen in patients with schistosomiasis, is suspected to be caused by antimonials. Life-time studies in mice, who were fed 5 ppm antimony as antimony potassium tartrate in drinking water, showed no evidence of cancer induction. However, a slightly higher risk of lung cancer was observed among antimony workers. The effect of smoking habits or other factors were not evaluated in these studies.

Arsenic

The role of arsenic in carcinogenesis is controversial. The IARC believes that there is sufficient evidence to show that inorganic arsenic compounds are skin and lung carcinogens in humans. The risk, however, appears to be marginal. A WHO study estimated 1% increase in lung-cancer incidence in workers exposed for at least 25 years to 0.25$\mu g/m^3$ of non-soluble inorganic arsenic, and for the general population, 0.75% per 1$\mu g/m^3$.

Serious doubts have been raised in recent years about the alleged carcinogenic effects of arsenic. In epidemiological studies linking occupational exposures with lung cancer, the populations were probably not exposed to arsenic compounds only, but to a multitude of carcinogens. Most animal experiments failed to induce cancer. Rather some reports indicate reduction in the frequency of lung or breast cancer following long-term feeding of arsenicals in experimental animals. People getting arsenical water in USA had a better health record than those getting non-arsenical water. It was further pointed out that if inhalation of arsenic trioxide really caused lung cancer, the disease would have reached epidemic proportions in Prague, where the level of arsenic in air from coal burning is very high. The current opinion is that arsenicals, like many other nutrients (e.g. vitamin A, selenium, zinc etc.), may have synergistic, antagonistic or no effect on carcinogenesis depending upon the form, dosage, and method of administration. The metal may be considered as a promoter, rather than the initiator, of the cancer process. At present, the state-of-the-art knowledge about the carcinogenic association of arsenic is not wholly understood and warrants further scrutiny.

Beryllium

Experimental and epidemiological studies have shown that beryllium can cause lung cancer, granulomas and osteogenic sarcomas. It selectively binds to regulatory non-histone nuclear proteins, inhibits phosphatase enzymes, interferes with gene transcription and inhibits DNA synthesis. The genotoxic effects were shown in various mammalian cell assay systems. Species differences for the induction of pulmonary tumours were also observed (sensitive : rat, human, resistant : hamster).

Cadmium

Cadmium exposure has been associated with higher incidence of lung cancer. Data pertaining to stomach cancer is conflicting while the incidence of prostate cancer could not be related to its exposure. Oral uptake of the metal does not lead to tumour induction while its inhalation leads to carcinogenesis. About 1 μg Cd/m^3 in air inhaled corresponds to about 60 $\mu g/g$ of the metal in lungs and leads to 50% incidence of pulmonary tumours. Smoking cigarettes significantly enhances the total body-burden of cadmium (upto 100 ng per cigarette) with a proportionate increase in the risk of cancer. Selenium and zinc may have protective effects against cadmium-induced carcinogenesis.

Chromium

Chromium elicits mutagenic, genotoxic and carcinogenic effects. Its hexavalent compounds, being actively transported into cells, are more potent carcinogens than the trivalent salts, which can enter the cells only to a limited extent. Besides, chromium (VI) has been shown to react with H_2O_2 to produce reactive oxygen species, which may further augment the toxic response. Workers handling chromate pigments and employed in chromate-manufacturing factories are at a higher risk of developing cancers of the nasal septum, lungs and skin. Experimental studies conducted for the induction of cancer in animal models employing respiratory or parenteral administration of chromium compounds, have not yielded reliable and unequivocal results.

Cobalt

Studies in bacterial test systems, cultures of human leucocytes and experimental animals show a reduction in the mitotic index. Cobalt sulphide was shown to be a more potent carcinogen than its oxide in implantation experiments in mice. There is some controversy regarding the goiterogenic effect of cobalt chloride, which is some times used to combat anaemia. Reports of higher incidence of bronchial carcinoma among cobalt miners are also available.

Copper

Copper sulphate has been shown to induce cutaneous cancer in mice and testicular teratomas in fowl. Cu (II) causes depurination of DNA, an event which may lead to mutagenesis. Cutaneous application of a carcinogen results in the epidermal loss of copper. A high incidence of lung cancer has also been reported in coppersmiths.

Biochemical studies reveal significantly higher copper concentrations in serum/plasma of patients with bronchial carcinoma, leukaemias, carcinoma of the breast, sarcomas/osteosarcomas, and Hodgkin's disease. However, a clear cause—effect relationship has not been established.

Gold

Gold is not carcinogenic. Certain gold compounds exhibit antipemphigus, cytotoxic and anti-tumour effects. Radioactive gold (^{198}Au) has been used in the form of a colloidal solution for radiotherapy in cancers of the gastro-intestinal tract, skin, connective tissues, bladder, pleura, ovaries and prostate and pituitary glands. Such therapy, however, has limited utility and manifests many side-effects.

Iron

Iron is suspected to be carcinogenic. Fats heated upto 300°C in iron vessels or in glass vessels in the presence of iron fillings also acquire this property. Painting the carcinogen on the skin of experimental animals leads to loss in epidermal iron within 10 days (up to 50%) and its continued application causes a further loss. In liver tumours, a considerable reduction in iron-containing enzymes such as cytochrome P-450 and catalase is noted. Self-medication with iron-containing tonics by middle-aged persons has been reported to cause a number of deaths from intestinal and stomach cancer. An enhanced mortality due to increased lung cancer has been reported following occupational exposure to iron dusts. Besides, iron particles probably serve as carriers for other carcinogens. It has also been observed that a population with iron overload is at higher risk of cancer incidence. The risk is further aggravated in smokers.

Lead

Lead compounds are reported to be mutagenic and genotoxic, causing disturbance in enzymatic functions and inducing chromosomal aberrations. Until early 1960s, most lead studies gave negative results regarding its carcinogenicity. However, it was shown later that acetate, subacetate, and phosphate of lead, when injected or administered through diet, could induce renal tumours in rodents. Kidney tumours develop at very high doses of lead and only in animals with obvious symptoms of acute lead toxicity with the appearance of nuclear inclusion bodies in renal cells. Some human studies reveal an increased incidence of cancer-mortality (gastro-intestinal tract, respiratory and renal cancers) among lead smelter and battery plant workers. However, only long-term exposures have been linked to the induction of cancer. The IARC and US EPA have recommended further investigations on these aspects for conclusive data. The metal may be considered as a carcinogen or a tumour promoter for human subjects.

Manganese

Studies in micro-organisms give some indication regarding the mutagenic actions of manganese. Its mutagenic or carcinogenic effects have not been confirmed in mammals. Some reports, on the contrary, show higher death rates from cancer in areas where its content in drinking water is low.

Mercury

Mercury has been shown to elicit antimitotic and cytotoxic effects and induce aneuploidy or hyperploidy and chromosomal breaks/aberrations in cells. Mice, who were fed 15 ppm methyl mercury chloride in diet, developed renal tumours. Sarcomas produced in rats were in areas of direct contact with the metal following its intra-peritoneal injection. Human studies showing association of mercury exposures with cancer are not available.

Molybdenum

In a statistical study, air-borne concentrations of molybdenum were correlated with the incidence of mortality from lung cancer, bronchitis and pneumonia. Other reports do not provide any evidence to show carcinogenic links of molybdenum.

Nickel

Experimental and epidemiological evidence indicates nickel to be mutagenic and carcinogenic in both animal models and man. Administration of soluble salts in drinking water promotes the carcinogenic action of nitrosamine in rats. The metal has been related to the incidence of malignant tumours of nasal cavity, larynx, sinuses and lungs in nickel-refinery workers exposed to its dust and nickel-carbonyl vapours. The average latent period was reported to be 22 years (range : 6-30 years). A correlation was also shown between nickel concentrations in town-water supplies in Iowa (USA) and the incidence of lung and bladder cancers. Higher amounts of nickel were reported in food chain in areas with high incidence of cancer as compared to other rural areas in Poland.

Platinum

Cisplatinum and carbaplatinum compounds have been used in cancer therapy.

Plutonium

^{239}Pu is a bone-seeking radioactive carcinogen. Malignant fibrosarcomas have been produced in mice with as low concentration as 1 μg.

Radium

The human body cannot excrete radium. The metal, which is deposited in tissues, particularly in bones, can be removed only by slow decomposition (biological half-life : 1620 years). Bone tumours follow ingestion or parenteral administration of radium. It is an occupational hazard to workers employed in gama radiography and luminous dial-painting industries. Patients with osteogenic sarcomas exhibit a total body burden of upto 50 μg of radium. The metal has also been linked to the occurrence of leukaemias. The isotope ^{226}Ra (in the form of sealed-in tubes, needles, plaques, moulds, etc.) finds clinical use for irradiation in carcinomas of buccal mucosa, cheeks, tongue, larynx, anal canal, uterus, vagina, eyes and ears.

Selenium

Selenium is a cancer-protecting agent. It retards the growth of certain chemically induced tumours. Epidemiological studies reveal a lower incidence of cancer in areas providing adequate dietary intake of selenium. In cancer clusters and in areas with high incidence of leukaemias, on the other hand, the levels of selenium in drinking water were found to be extremely low. Furthermore, a negative correlation was noted between blood selenium concentrations and cancer mortality. The mean values for human blood selenium levels in subjects from USA and New Zealand ranged between 0.08 and 0.2 ppm. These populations showed high incidence of leukaemias (6.6-7.0 per 100,000) and mortality from breast cancer (16.9-23.3 per 100,000). In Venezuela, which has soils rich in selenium, the inhabitants showed high blood-selenium levels (0.81 ppm), low incidence of leukaemias (2.8 per 100,000) and low mortality rate for breast cancer (2.6 per 100,000). Patients with cancers of stomach, colon, liver, pancreas, ovaries and Hodgkin's disease also showed significantly low values for blood/serum selenium concentrations.

Some contradictory reports showing the carcinogenic effects of selenium in experimental animals are also available. However, the metal does not induce cancer at low or moderate doses.

Silver

Colloidal silver is carcinogenic. Subcutaneous implantation of silver foil/particles has been shown to induce sarcomas in rodents. The metal does not seem to pose a risk to human beings through contact or ordinary use.

Strontium

Experimental induction of osteogenic sarcomas has been demonstrated in mice with ^{90}Sr. Radioactive ^{89}Sr and ^{90}Sr, on the other hand, have also been used for symptomatic relief in bone sarcomas, eyelid melanoma and mycosis fungoides.

Thallium

Thallium has been reported to cause DNA breaks, mutagenesis and carcinogenesis in experimental animals.

Thorium

Injected thorium stays life-long in body tissues and is suspected to cause bone sarcoma and leukaemias. Its use as a contrast media in X-ray photography should, therefore, be considered as an occupational hazard.

Tin

There are conflicting reports regarding the carcinogenicity of inorganic and organic tin compounds. A tremendous increase (400-600%) in blood-tin levels was observed after tumour (e.g., breast cancer) transplantation. Its concentration in cancerous tissue (breast cancer : 0.72 μg/g) was, however, found to be significantly lower than that in normal tissue (healthy breast :2.36 μg/g).

Uranium

Uranium possesses both mutagenic and carcinogenic effects. With the advent of atomic energy programmes, increased production and use of uranium has become a health hazard. High incidence of lung cancer is reported among uranium miners.

Vanadium

Mice given a 5 mg/ml of vanadyl sulphate for a life-time develop cancer. There is no report to suggest any association between this metal and cancer in human beings.

Zinc

The carcinogenic properties of zinc have not been fully investigated. Like other chromates, zinc chromate is a carcinogen. Application of hydrocarbon carcinogens causes a decrease in skin-zinc content. The metal is needed for cellular proliferation of existing tumours. Tumour growth is, therefore, retarded in zinc deficiency. The metal does not pose a risk for carcinogenesis in humans. However, significantly low concentrations of zinc in blood, plasma or serum have been reported in bronchial carcinoma, Hodgkin's disease, lymphoma, multiple myeloma and chronic lymphocytic leukemia. In cases of carcinoma of prostate glands, the organ contained low zinc concentrations but the values were high in hyperplastic prostate glands. Significantly high zinc levels were found in the liver of carcinoma patients. No comparable increase in the zinc content of kidney, spleen, heart, or pancreas was observed in autopsy samples of these patients. The significance of these findings so far remains elusive.

Colloidal radioactive ^{63}Zn fluid has been used for irradiation of serous cavities, lungs, bladder or for interstitial injection in tissues. This is particularly useful in malignant effusions of the pleural and peritoneal cavities requiring paracentesis. The metal also finds use in chemosurgery of malignant skin tumors.

Zirconium

Axillary granuloma can result from cosmetics that contain zirconium salts added as antiperspirant and deodorant. The use of these salts has, therefore, been discouraged.

4.6.3 Teratogenic Effects

Teratogenicity is related to sub-lethal fetotoxicity rather than to genotoxic effects. It is not heritable as the genetic material remains unaffected in most cases. Exposure to metals during the development of the embryo may cause malformations in the unborn child. Studies in experimental and domestic animals suggest that the following metals/metalloids may have a teratogenic potential : arsenic, cadmium, chromium, copper, lead, manganese, mercury, molybdenum, nickel, selenium, thallium, thorium, and zinc. Though conclusive data linking most of these metals with birth defects in humans is not available, evidence from animal experiments suggests a careful watch against their being a possible hazard for mankind.

Arsenic

Inorganic arsenic compounds such as sodium arsenite and sodium arsenate have been shown to cross the placental barrier and elicit tetratogenic effects in mice and sea urchins. Sometimes, these effects are not site-specific, such as following the exposure of arsenic to mice, but, in other cases, the specific abnormalities have been recorded. These examples include urogenital defects in rats, neural-tube defects in hamsters and mitotic and developmental abnormalities in sea urchins. Organic arsenicals, generally, do not have teratogenic properties though some reports indicate occurrence of *spina bifida* with methyl arsenate in chicks. Hamster is the most sensitive species in this regard. Protective action of selenium and BAL against arsenic-induced teratogenicity has been reported in hamsters and mice. According to an IARC report, offsprings of arsenic-smelter workers show low birth-weight and malformations.

Cadmium

Cadmium compounds, when injected in high doses, induce teratogenicity in rodents. Women subjected to industrial exposure with this metal gave birth to underweight infants, but no birth defects could be observed except for the occurrence of a few cases of rickets.

Chromium

Cr (VI) and Cr (III) compounds have been shown to induce teratogenicity in mice, golden hamster and chicken. Chromium levels were detected in foetal and placental tissues (abortive material) obtained during the 12th week of pregnancy. Among occupationally-exposed women working at a bichromate plant, the metal concentrations were observed to be as high as 114 μg% in the foetal tissues and 13.5 μg% in placenta. The corresponding values in non-exposed women were 9.28 μg% and 3 μg % respectively. However, no teratogenic effects could be recorded.

Cobalt

Cobalt chloride administered in doses of upto 5 mg/kg during pregnancy, exhibits marked embryotoxic effects in rats. The major defects include reduction in the ossification zones of skeletal bones and foetal size. Pregnant women receiving 75-100 mg of cobalt chloride per day for the treatment of anaemia, showed no malformations in their babies.

Copper

Copper-induced teratogenic effects have been reported in hamsters.

Lead

Intragastric administration (0.005 mg/kg in drinking water throughout the gestation period) and chronic inhalatory exposure (for 22 weeks) to pregnant rats revealed penetration of lead compounds through the placental barrier with increased embryonic mortality, a reduced foetal size and an impaired post-natal development of the offsprings. Evidence of teratogenic effects could be observed in many experimental animals, including mice, rats, hamsters and rabbits. Reduced gestation and higher frequency of abortions are observed in exposed women. Some reports indicate increased occurrence of various minor foetal malformations related to pre-natal lead-exposure in women, but the evidence for such an association remains elusive. Studies using the Mental Development Index (MDI), however, link maternal blood-lead levels as low as 10-15 μg/dl with growth retardation, deficient neurobehavioural performance and impaired post-natal development in children.

Manganese

Teratogenic actions of manganese compounds have been reported in mice but not in human subjects.

Mercury

Methyl mercury easily crosses the placental barrier. The post-natal excretion is very slow. The suckling rats are reported to have 19-fold higher level of mercury in their brain as compared to adults. Inorganic mercury does not readily get transferred across the placenta. Inspite of these differences, both organic and inorganic forms attain significant levels in breast milk and thus may be toxic to infants.

Organic mercury compounds have been shown to produce embryotoxic and teratogenic effects in most species including mice, rats, hamsters, rabbits, domestic animals, and humans. The common defects as a result of organic mercury exposure include intra-uterine death, foetal resorption, still-births, reduced foetal size and birth weight, cleft palate, facial malformations, abnormal skull and teeth, missing limbs, encephaly, encephalocele, hydrocephalus, cataract, blindness, growth retardation, ataxia, cerebral palsy, and neurobehavioural deficit. Some workers believe that ingestion of 1 kg of fish contaminated with methyl mercury (even within the legal limits) may produce significant behavioural deficit in human offsprings. The biochemical markers of methyl-mercury-exposure include elevated concentrations of orinthine decarboxylase in foetal brain and kidneys. Pathological changes have also been observed in the developing nervous system and renal tubules following its pre-natal or post-natal exposure. Epidemiological studies from Japan and Iraq revealed uneventful deliveries in exposed women. Abnormalities of skull and teeth, depressed general and mental development, decreased alertness and intelligence, ataxia and cerebral palsy may also develop later in infants. The damage to the foetal nervous system is usually permanent. For example, some degree of ataxia could be observed in surviving offsprings even after 30 years of the Minamata epidemic.

Molybdenum

Studies in rats showed that ingestion or inhalatory exposure to molybdenum compounds leads to a higher incidence of post-implantation foetal death, reduced birth-weight and inhibition of ossification in the foetus. Teratogenic effects have also been visualized in some domestic animals. There is no evidence of the occurrence of similar effects in humans.

Nickel

Teratogenic and embryotoxic effects of nickel compounds have been demonstrated in mice, rats, and hamsters. The metal elicits its adverse actions in the developing organism at relatively small doses in the early stages of pregnancy. Nickel chloride administered to pregnant rats caused foetal resorptions, increased embryonic mortality and foetal malformations. Exposure to nickel carbonyl in pregnant animals induced ocular abnormalities such as anophthalmia and microphthalmia in rat pups and exencephaly and cystic lungs in the hamster pups. Human placenta is permeable to nickel compounds, which is evident from an analysis of its content in the human embryo which shows its maximum accumulation between the 12th and 19th week of pregnancy. Reports regarding its effects on the new-born are not available.

Selenium

An elevated level of selenium reduces fertility and elicits embryotoxic and teratogenic effects in birds, rats and in some domestic animals. Inhalatory exposures to air containing > 0.1 mg Se/m^3 were shown to induce increased foetal mortality and disturbed ossification processes in rats. Birds are reported to be more sensitive to selenium-induced reduced fertility and foetal malformations. This usually occurs through ingestion of selenate and/or selenomethionine contained in edible plants, such as wheat. Though human data in this regard is inadequate, some reports suggest the possibility of such effects following its industrial exposures.

Thallium

Embryotoxic and teratogenic effects of thallium compounds have been reported in mice, rats, cats, and chicken. Some of the findings in experimental animals, however, could not be reproduced.

Thorium and Uranium

The metals emitting ionizing radiations at high doses have been shown to be teratogenic. Developmental defects and impaired brain-function in both animals and man have been observed.

Tin

Both the organic and inorganic tin compounds have been demonstrated to be without teratogenic properties. However, birth defects in rabbits following exposure to tricyclohexyltin miticides have been reported. Not much human data is available in this regard.

Vanadium

A high level of exposure to vanadium was shown to produce embryotoxic effects such as reduced litter size in experimental animals. Foetal malformations, however, were not observed. An analysis of human placental tissues from inhabitants of large industrial areas in Germany revealed a moderate to high tissue vanadium concentrations.

4.6.4 Mechanisms of Toxic Manifestations

The toxic response of a metal represents its incompatibility with living cells at a particular concentration/exposure level. While gradual concentration changes over a long period may lead to the development of susceptibility for its adaptation and tolerance, sudden elevations in its concentration often cannot be easily counteracted and usually result in toxic manifestations which may ultimately turn out to be fatal. This may also be true for essential metals at doses exceeding their maximum tolerance limits. The question as to how the toxic effects are elicited has no simple answer. The vast diversity of symptoms shown by various metals and their organic and inorganic compounds in different species, individuals and situations (health/disease, nutritional status, environmental factors, etc.) makes it extremely difficult to delineate a uniform mechanism. Though a general pattern does not exist, some common pathways for many metal-induced toxicities are discernible. The primary targets may often be one or more regulatory proteins/enzymes within cells. Enzymatic inhibitions and changes in the structure and function of bioproteins lead to impairment of vital functions. The generation of free radicals, with consequent damage to cell and organelle's membranes, is another frequently observed mechanism. Physical damage by particulate size, changes in the redox potential, inherent irritant/corrosive properties, allergic sensitization reactions, breaking of hydrogen bonds, damage to DNA and ATP, chromosomal aberrations, release of serotonin and other endogenous mediators, interactions with essential nutrients and genetic material, myelosuppression, immunosuppression etc., are some other ways through which metal/metalloids exert their toxic effects. Possible mechanisms/mediating factors of toxicity for individual metal/metlaloids are shown in Table 20.

TABLE 20 MECHANISMS OF METAL TOXICITY

METAL/METALLOID		MECHANISM(S) OF TOXICITY
Aluminium	(i)	Disturbances in calcium and ferritin concentrations in the brain.
	(ii)	Dephosphorylation of proteins, e.g., complex formation with inositol phosphate.
	(iii)	Activation of protein.
Arsenic	(i)	Binding to thiol anions.
	(ii)	Inhibition of enzymes involved in cellular respiration and SH-group-containing enzymes.
	(iii)	Interference with ATP synthesis by forming unstable arsenate esters.
	(iv)	Inhibition of haeme synthesis through the inhibition of haeme synthetase and the δ-aminolevulinic acid synthetase enzyme.
	(v)	Axons are more sensitive to the toxic effects of arsenic. The precise mechanisms causing axonal degeneration are not known.
Beryllium	(i)	Selective binding with high affinity to regulatory non-histone nuclear proteins.
	(ii)	Inhibition of enzymatic phosphorylation reactions.
	(iii)	Blocking the G-phase of the cell cycle and interference with DNA synthesis.
Cadmium	(i)	Displacement of zinc from critical site in zinc-containing enzymes.
	(ii)	Reaction with proteins and other biological molecules. High affinity for sulhydryl and imidazole nitrogen groups.
	(iii)	Interference with tubular resorption of low molecular-weight proteins.

METAL/METALLOID	MECHANISM(S) OF TOXICITY
	(iv) Non-metallothionein Cd and nuclear Cd are determinants of cadmium toxicity. Liberation of cadmium from a metallothionein-bound metal, through digestion by lysosomes, results in toxic effects.
Chromium	(i) Allergic and corrosive nature.
	(ii) Dermal and membranous absorption of immunologically active forms of chromium.
	(iii) Damage to DNA through interference with DNA-polymerase enzyme and free-radical formation.
	(iv) Carcinogenic action.
Cobalt	Myocardial toxicity is attributed to a decreased oxidation of pyruvate in the heart muscle.
Copper	(i) Local irritation of gastric mucosa.
	(ii) Impaired lysosomal excretion of hepatic copper in bile in Wilson's disease due to a genetic defect leads to its accumulation in liver with a consequent deficiency of ceruloplasmin in serum.
	(iii) Release of hepatic copper into the blood stream and extracellular fluids causes damage to susceptible cells including those of brain and liver.
Gold	Partial or complete myelosuppression.
Iron	(i) Catalytic effect on the production of the hydroxyl radical.
	(ii) Peroxidation of lysosomal membranes.
	(iii) Depolymerization of hyaluronate.
	(iv) Release of serotonin.

METAL/METALLOID		MECHANISM(S) OF TOXICITY
Lead	(i)	Interference with several enzymes involved in haeme synthesis, e.g. δ-amino-levulinic acid synthetase (ALA-S), δ-aminolevulinic acid dehydratase (ALA-D), ferrochelatase, etc. and reduction in haemoprotein biosynthesis.
	(ii)	Oxidative damage to the erythrocyte membrane.
	(iii)	Depression of serum dihydroxychole-calciferol, possibly through the inhibition of cytochrome P-450.
	(iv)	Involvement of calcium, chlinergic, monaminergic (dopamine, norepinephrine, 5-hydroxytryptamine) and the GABA-ergic system has been suggested as leading to neurotoxic effects of lead.
	(v)	Altered permeability of brain capillaries, segmental demyelination and axonal degeneration of peripheral motor nerves.
	(vi)	Blood pressure and renal effects are mediated through the renin-angiotensin system.
Manganese	(i)	Disturbances in the brain catecholamine metabolism.
	(ii)	Inhibition of dopamine, melanin and serotonin synthesis in the basal ganglia of brain.
Mercury	(i)	Affinity for thiol groups.
	(ii)	Neurotoxicities of $MeHg^+$ and Hg^o are distinctly different. The effects of mercury vapour on brain function are neuropsychiatric in nature, whereas those of $MeHg^+$ are largely sensorimotor. Inorganic Hg^{2+} is 100 times more potent than $MeHg^+$ as inhibitor of brain muscarinic

METAL/METALLOID		MECHANISM(S) OF TOXICITY
		receptors. Though Hg^{2+} has been shown to induce lipid peroxidation in isolated rat hepatocytes, it is not the sole cause of cellular toxicity. The latter is probably mediated through mechanisms affecting intracellular calcium concentrations.
	(iii)	Impairment of primary and secondary immune response.
Nickel	(i)	Nickel carbonyl inhibits ATP-ase and RNA-polymerase activities in target tissues.
	(ii)	Peroxidative degradation of membrane lipids.
	(iii)	Inhibition of natural killer-cell activity, T-cell mediated immune response and macrophage function.
	(iv)	Allergenic and carcinogenic action.
Platinum		Myelosuppression
Radium		Ionizing radiation
Selenium	(i)	Ability to block the actions of the SH-group of many enzymes and sulfur-containing amino acids.
	(ii)	Strong interactions with nutrients, e.g., vitamin C and E.
Silver	(i)	Local irritant/corrosive action.
	(ii)	Antagonism to essential nutrients, e.g. selenium, copper (reduced ceruloplasmin synthesis), vitamin B_{12} and vitamin E.
Thallium		Inhibition of enzymatic functions.
Thorium		Ionizing radiation.

METAL/METALLOID		MECHANISM(S) OF TOXICITY
Tin		Binding with mitochondria and inhibition of oxidative phosphorylation.
Uranium	(i)	Ionizing radiations.
	(ii)	Free UO_2^{2+} reacts with renal tubular cells.
Vanadium	(i)	Irritative action of vanadium fumes on the respiratory epithelium.
	(ii)	Release of histamine is responsible for its asthmogenic action.
	(iii)	Inhibtion of Na^+, K^+-ATPase and vasoconstrictive effects.
	(iv)	Reduced synthesis of co-enzyme A.
Zinc		Impaired immune responses; inhibitory effect on lymphocyte stimulation, chemotaxis and phagacytosis of granulocytes.

CHAPTER 5

FUTURE TRENDS

The growth homeostasis is linked to a large number of chemical elements which have complex interrelationship with each other. The growing industrial activity to satisfy economic drives and human comforts is associated with large-scale contamination of the biosphere by various pollutants including poorly biodegradable heavy metals. Can life withstand indefinitely such drastic changes in our environment ? Fortunately, some awareness about the seriousness of the problem is now discernible. These aspects have received the global attention of scientists in recent years. Multidisciplinary research projects are underway worldwide to evaluate the various facets of health hazards posed by increasing exposures to these metals. These researches also include deficiency-related problems, which sometimes arise due to regional influences but are often caused by malnutrition.

Albeit many questions remain unanswered and an understanding of precise mechanisms involved is so far elusive. The future trends of research in environmental aspects of heavy metals will, therefore, be obviously directed to solve many of these difficult problems which include :

* Routing and incorporation of metals into regulatory and catalytic proteins. Identification of precise steps and pathways involved in these processes.

* Unravelling mechanism of action of various metalloproteinases particularly in relation to the malignant transformation.

* Identifying metal switches in the DNA of organisms and their role in the manifestation of heavy-metal toxicity.

* Discovering the diverse strategies adopted by organisms to absorb, store and utilize essential metals while, at the same time, protecting themselves against metal toxicity.

* Identifying the mechanisms of detoxification of heavy metals and regulation of detoxifying elements.

* Understanding the mechanisms involved in target-organ toxicity of heavy metals, emphasizing their involvement in teratogenic, neurotoxic, carcinogenic and mutagenic effects.

* Discovering essential functions and therapeutic effects of heavy metals.

* Identifying zinc finger domains and other similar metalloproteins which transcriptionally regulate gene expression. Possible consequences of replacement of zinc by toxic metals from these sites.

CHAPTER 6

SUGGESTED READINGS

Bowen, H.J.M., *Environmental Chemistry of the Elements*, Academic Press, New York, 1979.

Bowman, M.C.: *Handbook of Carcinogenesis and Hazardous Substances*, Marcel Dekker Inc., New York, 1982.

Brown, S.S. and Kodama, Y. (editors), *Toxicology of Metals, Clinical and Experimental Research*, Ellis Horwood Ltd, Chichester, 1986.

Carson, B.L., *Toxicology and Biological Monitoring of Metals in Humans*, Lewis, Michigan, 1986.

Comar, C.L. and Bronner, P. *Mineral Metabolism : An Advanced Treatise*, Vol. 2, Part A, Academic Press, New York, 1964.

De Koning, H.W. (editor) : *Setting Environmental Standards : Guidelines for Decision Making*, UNEP/WHO, B.R. Publishing Corporation, Delhi, 1987.

Ellenhorn, M.J. and Barceloux, D.G., *Medical Toxicology : Diagnosis and Treatment of Human Poisoning*, Elsevier Science Publishing Co. Inc., New York, 1988.

Farrell, N. *Transition Metal Complexes as Drugs and Chemotherapeutic Agents*, Kluwer Academic Publishers, Dordrecht, 1989.

Frieden, E. (editor), *Biochemistry of the Essential Ultra Trace Elements*, Plenum Press, New York, 1984.

Goyer, R.A. and Mehlaman, M.A. (editors), *Toxicology of Trace Elements*, John Wiley and Sons, New York, 1977.

Greger, J.L., Aluminium and Tin, *Wld. Rev. Nutr. Diet*, Karger, Basel **54**, 255-285, 1987.

Gupta, P.K. and Salunkhe, D.K., *Modern Toxicology*, Vol. 2. *The Adverse Effects of Xenobiotics*, Metropolitan Book Co. Pvt. Ltd., New Delhi, 1985.

Hamdard, M.E. and Vohora S.B., *Medical Elementology: Practical Applications*, Institute of History of Medicine and Medical Research, New Delhi, 1989.

Iyengar, G.V., Kollmer, W.E. and Bowen, H.J.M. (editors), *The Elemental Composition of Human Tissues and Body Fluids*, Verlag Chemie, New York, 1978.

Izmerov, N.F. (editor: *Scientific Reviews of Soviet Literature on Toxicity and Hazards of Chemicals*, report nos. 11 (mercury), 17 (thallium), 20 (arsenic), 42 (lead), 58 (nickel), 63 (beryllium), 67 (vanadium), 68 (chromium), 69 (cadmium), 71 (antimony), 78 (selenium), 94 (zinc), 100 (cobalt) and 105 (molybdenum), UNEP/ IRPTC, Centre for International Projects, GKNT, Moscow 1982-1988.

Kabata-Pendias, A. and Pendias, H, *Trace Elements in Soils and Plants*, CRC Press Inc., Boca Raton, Florida, 1984.

Kumar, R. (editor), *Environmental Pollution and Health Hazards in India*, Ashish Publishing House, New Delhi, 1987.

Lansdown, R. and Yule, W. (editors), *The Lead Debate*, Croom Helm, London, 1986.

Lewis, A.J. and Walz, D.T., Immunopharmacology of gold, *Prog. Med. Chem.* **19**, 1-58, 1988.

Merian, E. (editor), *Metals and Their Compounds in the Environment : Occurrence Analysis and Biological Relevance*, VCH, Wienheim, 1991.

Moran, J.M., Morgan, M.D. and Wiersma, J.H., *Introduction to Environmental Science*, Ed. 2, W.H. Freeman and Co., New York, 1986.

Moore, J.W. and Ramamurthy, S., *Heavy Metals in Natural Waters*, Springer Verlag, New York, 1984.

Nath, R., *Environmental Pollution of Cadmium: Biological, Physiological and Health Effects*, Interprint, New Delhi, 1986.

Nriagu, J.O., Global metal pollution : poisoning the biosphere ? *Environment*, **32(7)**, 7-33, 1990.

Prasad, A.S., *Trace Elements and Iron in Human Metabolism*, Plenum Medical Book Co., New York, 1978.

Prasad, A.S. and Oberleas, D., *Trace Elements*, Vol. 1, *Zinc and Copper*, Academic Press, New York, 1976.

Schroeder, H.A. : *The Trace Elements and Nutrition : Some Positive and Negative Aspects*, Faber and Faber, London, 1976.

Siegel, H. (editor), *Metal Ions in Biological Systems*, *Vol. 20, Concepts on Metal Ion Toxicity*, Marcel Dekker Inc., New York, 1986.

Underwood, E.J., *Trace Elements in Human and Animal Nutrition*, Ed. 4, Academic Press, New York, 1977.

UNEP/ILO/WHO, *Environmental Health Criteria Series*, Report nos. 1 (mercury), 3 (lead), 15 (tin), 17 (manganese) and 18 (arsenic), World Health Organization, Geneva, 1976-1981.

Vercruysse, A. (editor), *Hazardous Metals in Human Toxicology*, Part B, Elsevier Science Publishers BV, Amsterdam, 1984.

Vohora, S.B. and Dobrowolski, J.W. (editors), *New Horizons of Health Aspects of Elements*, An Indo-Polish Book, Jamia Hamdard, New Delhi, 1990.

WHO, *Guidelines for Drinking Water Quality*, Vols. 1 and 2, World Health Organization, Geneva, 1984.

INDEX